POLITICS IN THE PORTUGUESE EMPIRE

POLITICS IN THE PORTUGUESE EMPIRE

The State, Industry, and Cotton, 1926–1974

M. ANNE PITCHER

CLARENDON PRESS · OXFORD
1993

Oxford University Press, Walton Street, Oxford OX2 6DP
Oxford New York Toronto
Delhi Bombay Calcutta Madras Karachi
Kuala Lumpur Singapore Hong Kong Tokyo
Nairobi Dar es Salaam Cape Town
Melbourne Auckland Madrid
and associated companies in
Berlin Ibadan

Oxford is a trade mark of Oxford University Press

Published in the United States
by Oxford University Press Inc., New York

British Library Cataloguing in Publication Data
Data available

Library of Congress Cataloging in Publication Data
Pitcher, M. Anne.
Politics in the Portuguese Empire: the State, industry, and cotton, 1926–1974 / M. Anne Pitcher.
p. cm.
Includes bibliographical references and index.
1. Cotton textile industry—Portugal—History—20th century.
2. Cotton growing—Portugal—Colonies—History—20th century.
3. Portugal—Colonies—Africa—Administration. I. Title.
HD9885.P82P58 1993 92-41457
338.4'7677'0091712469—dc20
ISBN 0–19–827373–8

1 3 5 7 9 10 8 6 4 2

Typeset by Best-set Typesetter Ltd., Hong Kong
Printed in Great Britain
on acid-free paper by
Bookcraft (Bath) Ltd., Midsomer Norton, Avon

Acknowledgements

The research and writing of this book has been not only challenging, interesting, and even fun; but also laborious and frustrating at times. To my friends and colleagues—who significantly increased the former and substantially reduced the latter—I am deeply grateful. The work originated as a D.Phil. dissertation at Oxford University and I owe an intellectual debt to my supervisor, Dr David Goldey, for his critical comments on the text and for reminding me that writing is artisanal. I would like to thank my thesis examiners, Professor David Birmingham and Dr Herminio Martins, for their stimulating discussion and useful suggestions for publication.

I should like to express my gratitude for the valuable assistance offered by the librarians at Queen Elizabeth House, Rhodes House, and the Bodleian Library, Oxford; Case Library at Colgate University; the ICS library and CIDAC library in Lisbon; the Biblioteca Nacional in Lisbon, and the Biblioteca Pública Municipal in Oporto. Dr Manuel Villaverde Cabral, when he was the Director of the Biblioteca Nacional, also provided useful introductions and advice. For their insights and help in directing me towards new lines of enquiry, I would like to thank Mr Gavin Williams, Dr Edgar Rocha, Dr Maria Filomena Mónica, Dr Manuel de Lucena, Dr José Brandão de Brito, Dr Jaime Reis, and especially Dr Gervase Clarence-Smith. The work of Dr Clarence-Smith has greatly inspired me and his extensive comments on my manuscript have been invaluable. Members of the Political Science Department at Colgate University also read portions of the manuscript and I am grateful for their suggestions. I am also indebted to Manuel José Ferreira de Sousa of the textile trade union in Guimarães. Lastly, I would like to thank Bibiana and Henrique Dantas for helping me to understand Portugal and its textile industry.

The book would never have been completed without the generous financial assistance provided by St Hilda's College, which awarded me an Overseas Bursary from 1985–7 and gave me a travel grant from the Muriel D. Wise Travel Fund in 1986. An Overseas Research Award given by the Committee of Vice-Chancellors and Principals of the United Kingdom also provided financial support for 1985–6 and I would like to thank the Office of Graduate Studies, Oxford University, for giving me a travel grant in 1986 and 1987. A Fulbright Award in 1988–9 to conduct research on the Portuguese textile industry and European integration gave me an additional year to understand Portuguese politics and industry. When the book was near completion, a research grant by the Colgate Research Council and financial support by the Dean of the Faculty at Colgate University funded additional research in Portugal in the summer of 1991. I would like to thank the chairman of my department, Dr Michael Johnston, for helping me to secure these funds and for providing two research assistants, Beth Turro and John Holzwarth, who helped me enormously in the final stages. For graciously offering an office for the summer at the Instituto de Ciências Sociais in Lisbon, I am obliged to Dr Filomena Mónica.

Finally, I am sincerely grateful to my friends and flatmates for patiently listening to my endless monologues about cotton prices, colonial markets, and textile firms. Doug Tilton gets a special mention for arranging the maps, and Kevin Rask, for preparing the graphs. I offer many thanks also to my parents for their constant encouragement. Most of all, I deeply appreciate the love, good humour, and emotional support of my husband, Warren Robert Shiell.

M.A.P.

Hamilton, New York
8 June 1992

Contents

Figures

Maps

Tables

Abbreviations

AIp	Associação Industrial Portuense Oporto Industrial Association
AIP	Associação Industrial Portuguesa Portuguese Industrial Association
BGC	*Boletim Geral das Colónias* *General Colonial Bulletin*
BGU	*Boletim Geral do Ultramar* *General Overseas Bulletin*
BNU	Banco Nacional Ultramarino Overseas National Bank
Bol. da DGI	*Boletim da Direcção Geral da Indústria* Bulletin of the Directorate-General of Industry
BPA	Banco Português do Atlantico Portuguese Bank of the Atlantic
CADC	Centro Académico da Democracia Christã Academic Centre of Christian Democracy
CAM	Companhia dos Algodões de Moçambique Mozambique Cotton Company
CICA	Centro de Investigação Científica Algodoeira Cotton Research Centre
c.i.f.	cost, insurance, freight
Cotonang	Companhia Geral dos Algodões de Angola General Cotton Company of Angola
CRCAR	Comissão Reguladora do Comércio de Algodão em Rama Cotton Regulatory Commission
CUF	Companhia União Fabril Union Manufacturing Company
DGI	Direcção Geral da Indústria Directorate-General of Industry
DGSI	Direcção Geral dos Serviços Industriais Directorate-General of Industrial Services

DSA	Direcção dos Serviços da Agricultura Directorate of Agricultural Services
ECGR	*Empire Cotton Growing Review*
EFTA	European Free Trade Association
ETN	Estatuto Trabalho Nacional National Labour Statute
FNPT	Federação Nacional dos Produtores de Trigo National Wheat Producers Federation
f.o.b.	free on board
FPLN	Frente Patriótica de Liberatação Nacional Patriotic National Liberation Front
FRELIMO	Frente de Liberatação de Moçambique Front for the Liberation of Mozambique
GEDES	Gabinete de Estudo de Divulgação Económica e Social Economic and Social Research Council
GNIAR	Grémio Nacional dos Importadores de Algodão em Rama National Guild of Cotton Importers
INE	Instituto Nacional de Estatística National Institute of Statistics
JEAC	Junta de Exportação de Algodão Colonial Colonial Cotton Export Board
MANU	Mozambique African National Union
MPLA	Movimento Popular de Liberatação de Angola Popular Movement for the Liberation of Angola
MUD	Movimento do Unidade Democrática Movement of Democratic Unity
Munaf	Movimento de Unidade Nacional Anti-Fascista Movement of National Anti-Fascist Unity
NATO	North Atlantic Treaty Organization
OEEC	Organization of the European Economic Community
OTN	*O Trabalho Nacional* *Journal of National Labour*

SAGAL	Sociedade Agrícola Algodoeira Cotton Agricultural Society
SNI	Secretariado Nacional da Informação National Secretariat of Information
SPN	Secretariado da Propaganda Nacional Secretariat of National Propaganda
UIE	União dos Interesses Económicos Union of Economic Interests

Note: The Portuguese use $ for their currency, e.g. 4$50 = 4 escudos, fifty centavos. For exchange rates, see Tables A1 and A2.

A Note on Sources and Materials

Although the authoritarian regime was overthrown in 1974, some archives in Portugal remain closed and many people associated with the old system refuse or are reluctant to discuss their involvement, particularly in areas considered politically sensitive such as the relationship between the textile industry, the state, and the colonies—the subject of this study. Establishing contacts with heads of industry or former government officials often requires years of letter writing, personal introductions, the presentation of credentials, and the building of confidence and trust. In my research, young executives in the textile industry were particularly helpful for the period immediately preceding the 1974 coup and the present. They were willing to talk about the current challenges faced by textile manufacturers, yet were ignorant of events between 1926 and 1960—precisely the period on which information is needed.

In order to present the interests and opinions of textile manufacturers on regime policy, colonial cotton, markets, and competition during the period between 1926 and 1974, this study makes use of numerous company reports. Because the authoritarian regime in Portugal actively enforced censorship, criticisms of policies tended to be surreptitious and veiled. Active disagreement with government policy was rare, but company reports in Portugal are very expansive—often discussing general political and economic issues in addition to their own financial position. As a result, they often give insight into the worries and troubles of the industry. I rely on these to illustrate points or to substantiate material found in other sources. Unfortunately, large rather than small companies were more likely to publish company statements, so the situations and points of view of small companies have been garnered from independent reports and analyses. The regime appears to have relaxed a bit in the 1950s and several studies more openly address the plight of small business. These studies also more critically examine the problems of the industry and give an indication of the government's and the colonies' contribution to them.[1]

[1] See e.g. the report by the Segundo Congresso da Indústria Portuguesa; works by Armando Carneiro, 1957–9; the thorough study by Maria Helena Garcia da

Government legislation and reports by corporate organizations involved with the textile industry and colonial cotton production tend to give the government's perspective. Portuguese practice is to give a brief description of the circumstances or problems addressed by legislation, and preambles to laws often give very detailed reasons for a policy and its purpose.[2]

For the colonies, in addition to some very thorough secondary sources, I use reports written by former members or departments of the colonial administration, colonial societies, conferences, and corporate organizations. Cotton concessionary company reports were helpful for illustrating points made in secondary sources or difficulties cited in the preambles to laws or other sources.[3]

Statistics for the metropole and the colonies are quite good for the post-1935 period, numerous but contradictory for earlier dates. I have tried to use the most informed sources and to verify them where possible. The Mozambique statistical record is complicated by the fact that the old-style concessionary companies located there (one of which lasted until 1942) often published their own statistics; their holdings, imports, and exports were not always included in official statistics on the colony. Therefore, many sources have gaps and are incomplete. Where appropriate, this is pointed out in the text and appendix.

Furthermore, from 1881 until the late 1950s there was no complete census of Portuguese industry. Until the 1957–9 census, the 1881 census was the last complete census according to the Instituto Nacional de Estatística (INE). The census in 1890 met with a lack

Fonseca; José Joaquim Teixeira Ribeiro, *Reorganização da indústria têxtil algodoeira* (Projecto de Relatório da Comissão Reorganízadora, 1966); Lino Lima, *A crise da indústria têxtil algodoeira* (Vila Nova de Famalicão: Centro Gráfico, 1967); and the report by the Comissão de Estudo da Indústria Têxtil, 1968.

[2] See e.g. Portugal, *Diário do Governo*, Série I, 1951–74; and the Ministério da Economia, *Boletim da Direcção Geral da Indústria*, Publicação Semanal, 1937–49; later changed to the *Boletim da Direcção Geral dos Serviços Industriais*, 1949–65, all under the Government and Corporate Publications and Official Documents section of the Bibliography.

[3] See Ismael Alves da Costa, 'A cultura de algodão em Moçambique', III Congresso Colonial Nacional, 1930; works by H. P. Taveira and the Sociedade de Geografia de Lisboa; books by E. K. de Queiroz Ribeiro; material of the Junta de Exportação de Algodão Colonial (JEAC); Mozambique, Direcção dos Serviços de Agricultura; Portugal, Agência Geral das Colónias, *Boletim Geral das Colónias*, 1946–51, and Agência Geral do Ultramar, *Boletim Geral do Ultramar*, 1952–69; and the colonial cotton concessionary companies cited in the Private Sector section of the Bibliography.

of co-operation from industrialists and its accuracy is questionable. Attempts to conduct a census in 1907, 1909, 1924, 1931, and 1950 never materialized.

Until the creation of a government body to regulate the textile sector in the late 1930s, statistics on industrial machinery and the numbers of textile mills during the Republican period and the early part of the *Estado Novo* were based on estimates. Therefore great leaps in the numbers of machines do not necessarily mean that production increased, but simply that the previous estimate was too low. This will be discussed in more detail in Chapter 2. When the Comissão Reguladora do Comércio de Algodão em Rama (CRCAR) was established in 1937 to organize cotton imports and regulate the textile industry, the statistics improved. For most of the information on the structure, imports, exports, and capital of the textile industry from 1937 to 1974, I rely on statistics furnished by the CRCAR, their representatives, and the government statistical department, the INE.

Introduction

At the intersection of an autarkic colonial policy and a nationalistic ideology, the Portuguese textile industry should have been a major beneficiary of nearly fifty years of authoritarian rule. After a military coup toppled the Republican government in 1926, the new regime initiated protectionist legislation designed to secure the colonial market for Portuguese textile exports and to promote colonial cotton production for use by the industry. Following the creation of the *Estado Novo* in 1932, industrial regulations and government agencies sought to control growth within the textile industry and limit competition amongst mills. In the 1940s and 1950s, modernization legislation, a commitment to development, and low colonial cotton prices were intended to make the industry more competitive abroad. Trade agreements, the supply of credit, and the entry of foreign capital into Portugal in the 1960s aimed at expansion into European markets. Indeed many of the regime's industrial and colonial policies specifically addressed the needs and interests of the Portuguese textile industry.

Yet by 1974, the year of the regime's collapse, most of the industry was in a shambles. Many of the mills were dilapidated and produced poorly made materials which were uncompetitive in the rest of Europe. Other mills, having modernized in order to compete effectively in markets abroad, had abandoned their old connection with the colonies because the cotton imported from there did not match in quantity or quality the needs of the industry. Colonial cotton, priced above that purchased on the international market, lay unsold on the dock either because industrialists could not afford to buy it or had decided to buy better cotton elsewhere. Labour was emigrating in droves from Portuguese mills to higher paying, more satisfying jobs in Northern Europe. A report on the state of the industry cited the poor conditions in mills, the unsuitable quality and quantity of colonial cotton supply, over-production, cutthroat competition, and the lack of sufficient markets for Portuguese textiles as some of the major elements of the crisis affecting the industry.[1] Ironically, these were the same conditions which had justified the implementation of industrial regulations, government agencies, protectionist policies, and the colonial cotton system nearly half a century earlier.

Why did autarky and intervention fail to operate in practice the way they were meant to in theory? Why did colonial and industrial policies, which were designed to benefit the industry, instead contribute to its ruin? To unravel these puzzles, this study critically examines the complex relationship which existed between the Portuguese state, the textile industry, and the empire in Africa from 1926 to 1974. It evaluates the costs and benefits of empire to Portugal's authoritarian regime and, more specifically, to the metropolitan textile industry. To assess the value of empire, the study broadly explores the role that Portugal's overseas possessions played in the achievement of the regime's political, economic, and ideological objectives. It discusses what and whose interests were served by the actions of the metropole in its colonies. The study documents in detail the creation and maintenance of the colonial cotton system and the acquisition of protected textile markets in Portugal's two largest African possessions, Angola and Mozambique. It analyses the motives behind colonial cotton policies, their actual operation in practice, and their impact on textile mills in the metropole.

The study also judges the effect of the authoritarian regime on the development of one of Portugal's major industries. Looking beyond the façade of harmony projected by the regime, it examines the articulation of interests by the textile industry and their representation or suppression by the Salazar and Caetano Governments. It investigates the purpose behind industrial regulations, government regulatory agencies, and development plans as well as their particular influence on textile manufacture. Finally, it situates the textile industry within the political and economic milieu which characterized the Portuguese authoritarian period in order to explain not only the industry's changing position in the hierarchy of state priorities but also the nature of the regime.

In exploring these issues, the study challenges the view propagated by the Salazar regime of a smooth and functional relation-

[1] Comissão de Estudo da Indústria Têxtil, *Relatório e projecto de regulamento de exercício da indústria têxtil*, Porto, 30 Apr. 1968 (hereinafter Comissão, *Relatório*).

ship between the state and the textile industry over colonial policy. Certainly, colonial cotton supply and textile markets were the pillars of Salazar's autarkic policies, yet claims by the state and its supporters about the benefits that the colonial connection afforded the metropolitan textile industry have often been accepted at face value, without proper substantiation, even by scholarly critics of Portuguese colonialism.[2] The position that industry gained from its association with empire rests on three assumptions: first, that colonial policy under Salazar was a rational system of exploitation designed to serve only metropolitan interests; secondly, that the outcome of the cotton policy was in fact beneficial to metropolitan interests; and thirdly, that in the case of the cotton scheme, the interests of domestic textile manufacturers coincided with those of the state. Many scholars tacitly adopt these assumptions when referring to the relationship between industry and empire; yet, ironically, they chronicle elsewhere in the former colonies a succession of irrational acts, failed and misguided policies, conflicts between the administration and companies over labour, and effective resistance by Africans against colonial policies.[3]

The contradiction could be resolved if scholars of Portuguese economic, political, and colonial history were to view the relationship between the Portuguese metropole and the colonies as an interactive process, a two-way exchange (however unequal) of interests and influences rather than a simple top-down imposition

[2] Among others, the following authors tend to give fairly brief and standard explanations of the cotton regime and its relationship to the Portuguese textile industry: Lawrence W. Henderson, *Angola* (Ithaca, NY: Cornell Univ. Press, 1979); Allen F. Isaacman and Barbara Isaacman, *Mozambique: From Colonialism to Revolution, 1900–1982* (Boulder, Colo.: Westview Press, 1983); Elisio Martins, *Colonialism and Imperialism in Mozambique: The Beginning of the End* (Copenhagen: Akademisk Forlag, 1975); Barry Munslow, *Mozambique: The Revolution and its Origins* (London: Longman, 1983); and M. D. D. Newitt, *Portugal in Africa: The Last Hundred Years* (London: C. Hurst & Co., 1981). David Hedges and Aurelio Rocha in their two articles, 'A reestruturação do colonialismo em Moçambique, 1938–1944', *Cadernos de História*, 5 (Feb. 1987), 41–73, and 'Moçambique durante o apogeu do colonialismo português, 1945–1961', *Cadernos de História*, 6 (Nov. 1987), 29–61, give a thorough account of the cotton regime but assume the interests of the state and the 'bourgeoisie' coincided in its creation and preservation.

[3] See authors above and the insightful articles by Allen F. Isaacman *et al.*, 'Cotton is the Mother of Poverty: Peasant Resistance Against Forced Cotton Cultivation in Mozambique, 1938–1961', *International Journal of African Historical Studies*, 13: 4 (1980), 581–615; and Allen F. Isaacman, 'Chiefs, Rural Differentiation and Peasant Protest: The Mozambican Forced Cotton Regime 1938–1961', *African Economic History*, 14 (1985), 15–56.

of metropolitan initiatives and objectives. Except for the work of Michel Cahen, Gervase Clarence-Smith, J. Vaz, and (for an earlier period) Gregory Pirio, the study of the interaction between the Portuguese metropole and the colonies has been largely neglected.[4] This is somewhat surprising, since a lengthy theoretical debate on the costs and benefits of empire for the metropole has dominated British and French colonial historiography.[5]

Located at the interface between the Portuguese metropole and the colonies, the textile industry provides an insightful case-study because it was an industry which relied on the colonies not only for supplies but also for markets. This placed it in the awkward position of depending on the colonies at both the beginning and the end of the production process. Such dependence made textile manufacturers uncomfortable about the association on the one hand, and on the other, gave cotton concessionary companies, African producers, the colonial administration, and Portuguese settlers a degree of political leverage they may not otherwise have had. In nearly fifty years of association with metropolitan industry, these colonial interests were far from passive; they frequently imposed economic constraints and political choices on the Portuguese state and the textile industry over questions of the quality, quantity, and price of cotton and the availability of textile markets. In the case of the Portuguese textile industry, therefore, the potential influence of empire on industry was enormous.

Focusing on the interplay between the colonies and the metropole invariably forces us to examine a major actor in the process—the

[4] For material on interaction between the Portuguese metropole and the colonies, Michel Cahen, 'Corporatisme et colonialisme: Approche du cas mozambicain, 1933–1979', Parts I and II, *Cahiers d'études africaines*, 23–4: 92–3 (1983), 383–417, and 5–24; José Manuel Vaz, 'Portugal and Her Overseas Territories: A Study in Economic Association and Co-operation', Ph.D. thesis (Exeter, 1978); Gervase Clarence-Smith, *The third Portuguese empire, 1825–1975: A study in economic imperialism* (Manchester: Manchester Univ. Press, 1985); Gregory Roger Pirio, 'Commerce, Industry and Empire: The Making of Modern Portuguese Colonialism in Angola and Mozambique, 1890–1914', Ph.D. thesis (UCLA, 1982).

[5] In the British and French colonial historiography, the debate has its origins in the Hobsonian–Leninist arguments on imperialism. For recent contributions to the debate, see e.g. P. K. O'Brien, 'European Economic Development: The Contribution of the Periphery', *Economic History Review*, 2nd ser., 25: 1 (Feb. 1982), 1–18; and P. J. Cain and A. G. Hopkins, 'Gentlemanly Capitalism and British Expansion Overseas II: New Imperialism, 1850–1945', *Economic History Review*, 2nd ser., 40: 1 (Feb. 1987), 1–26; and Jacques Marseille, *Empire coloniale et capitalisme française: Histoire d'un divorce* (Paris: Albin Michel, 1984).

Portuguese state. Study of the regime which existed from 1926 to 1974 has been marred for too long by historiographical prejudices and methodological difficulties. The historiography of Salazarism has concentrated on the political models espoused by ideologues of the *Estado Novo* and has not placed sufficient emphasis on the actual workings of the system. Until recently, much analysis of the nature and composition of Salazar's *Estado Novo* (which was legally constructed in 1933) and Caetano's *Estado Social* (which replaced it in 1968) was overly concerned with their institutional and ideological aspects and their proximity to an ideal theoretical notion of corporatism.[6] Studies of the corporative framework detailed extensively the institutions and regulations of the state, but few examined the interaction between specific sectors and the regime, or documented the implementation and effect of policies on interest groups or on the colonies. Those which did attempt to examine the influence or impact of groups on the state's behaviour and policies tended to be narrowly economistic and to focus mainly on the activities of the small but powerful financial-industrial conglomerates in Portugal.[7]

Much of the reason for this focus had to do with the nature of the state itself. Given an ideology which denied the existence of legitimate class or intersectoral conflicts and a security apparatus which suppressed them, it was difficult to study conflict or compromise in a society which was, by fiat, 'organic' and harmonious. Information on the inner workings of government was concealed and free comment discouraged; the nature of Portuguese education did not encourage students to study social and economic difficulties. As Lawrence Graham and Harry Makler pointed out in *Contemporary Portugal: The Revolution and its Antecedents*,

> Undoubtedly the humanistic emphasis of university education and strict censorship during the Salazar regime restricted the type of work produced,

[6] See e.g. Manuel de Lucena, *A evolução do sistema corporativo português*, i and ii (Lisbon: Perspectivas e Realidades, 1976); Walter C. Opello, *Portugal's Political Development* (Boulder, Colo.: Westview Press, 1985); and Howard J. Wiarda, *Corporatism and Development* (Amherst: Univ. of Massachusetts Press, 1977). The exception is Philippe C. Schmitter, *Corporatism and Public Policy in Authoritarian Portugal* (Beverly Hills, Calif.: Sage Publications, 1975) whose title reflects the scepticism with which he viewed the Portuguese variant. His epilogue, written after the coup of 1974, nearly rejects the term all together.

[7] Perry Anderson, 'Portugal and the End of Ultra-Colonialism', *New Left Review*, nos. 15–17 (1962), 83–102; and Francisco Rafael *et al.*, *Portugal: Capitalismo e o Estado Novo* (Lisbon: Edições Afrontamento, 1976).

and contributed to the fact that Portugal was invisible to much of the outside world. The aging dictatorship of António Salazar weathered the *Santa Maria* hijacking in 1961, and even revolutionary activities in its overseas territories were largely overlooked by the Western press and Western scholars.[8]

Manuel de Lucena's thorough study of the construction of Portuguese corporatism (on which much of the work in English was based) was researched and written in exile and necessarily depended largely on the Portuguese Constitution and government legislation for its sources.[9] According to Philippe Schmitter, even after the 1974 coup information on the activity of the state continued to be restricted as various political factions fought for access and control over the material.[10]

Recently, however, several studies have begun to examine the influence of groups on the nature of the *Estado Novo* and the formulation of policy, the function and behaviour of state-created institutions, the motivations behind government legislation, the impact of corporative ideology, and the conflict and co-operation between groups and individuals working within the government apparatus.[11] The following work charts the rise and fall of political challenges to the regime as well as shifts in its economic priorities in order to explain the state's changing role in the triangle of metropolitan, colonial, and textile interests. Again, the textile industry provides a good point from which to view the state because it was the subject of so much of the regime's protectionist, regulatory, and development legislation, and it was included in the corporate edifice constructed after 1933.

[8] L. S. Graham and Harry Makler (eds.), *Contemporary Portugal: The Revolution and its Antecedents* (Austin: Univ. of Texas Press, 1979), p. xv.

[9] Lucena, *A evolução*, i. 13.

[10] Philippe C. Schmitter, '"The Regime d'Exception" that became the Rule: Forty-eight Years of Authoritarian Domination in Portugal', in Graham and Makler (eds.), *Contemporary Portugal*, 3.

[11] See e.g. Fernando Rosas, *O Estado Novo nos anos trinta 1928–1938* (Lisbon: Editorial Estampa, 1986), for a discussion of the government's balancing act among the various industrial and agricultural groups in the 1930s; and his *Portugal entre a paz e a guerra* (Lisbon: Editorial Estampa, 1990), on the activity of the government during World War II; Manuel de Lucena's study of the *grémios da Lavoura* published in *Análise Social*; the work by José Brandão de Brito on the industrial regulation legislation; Alice Ingerson, 'Corporatism and Class Consciousness in Northwestern Portugal', Ph.D. thesis (Johns Hopkins, 1984); and studies of the various political factions in the *Estado Novo* by Manuel Braga da Cruz.

Focusing on a single industry cannot lead to a definitive assessment about the nature of the Portuguese state. Nevertheless an examination of the triangular relationship between the state, the colonies, and the textile industry exposes a regime which changed over time and over issues. In its infancy, the *Estado Novo* arbitrated amongst highly fragmented, but powerful interest groups in order to consolidate its power. As it matured, it became more openly interventionist and authoritarian. Concomitantly, depending on the interest group and the issue, the regime could be pragmatic, kleptocratic, autarkic, nationalistic, excessively fiscally conservative, and occasionally corporative. Such an assortment of guises suggests a certain responsiveness on the part of the regime, but as time passed, these actions revealed a lack of direction, a confused desperation.

The state's many responses invariably affected its relationship with textiles and the colonies. Policies sometimes contradicted each other; different goals of the regime also conflicted with the wishes of the textile industry. In the colonies, other priorities often overrode those of cotton concessionary companies and conflicts over labour, land, and prices affected output. Few participants experienced total satisfaction, least of all African cotton producers, who suffered enormously as a result of colonial agricultural policies.

Through a periodization which corresponds to major political and economic developments from 1926 to 1936, 1936 to 1946, 1946 to 1958, and 1958 to 1974, the study documents these changes, variations, and modifications in the relationships among the regime, the textile industry, and the empire in Africa. Chapters 1, 4, 7, and 10 on the authoritarian regime discuss the environment in which the government operated in order to provide a context for understanding its policies towards the textile industry and the colonies. These chapters, which correspond to the periods listed above, identify political challenges, economic strategies, sources of conflict, and dominant interests influencing the regime.

Chapters 2, 5, 8, and 11 on the textile industry examine the interaction between the state and industry. They look at specific legislation applied to the textile industry and the government organizations established to control it. They examine the possible interests and considerations involved in the framing of laws and those interests affected by legislation and state direction. Finally, these chapters trace the process whereby the colonies and the authoritarian regime affected the industry's growth and development.

Chapters 3, 6, 9, and 12 on colonial cotton production examine the implementation of cotton legislation and the creation of organizations to enforce it. They trace the numerous problems which arose over quantity, quality, and price of colonial cotton, and their effects on the textile industry. Chapter 12 which covers the final period from 1958 to 1974 discusses the colonial war and the events which finally led to the withdrawal of Portugal from Africa. A concluding chapter assesses the impact of nearly half a century of authoritarian rule on Portugal, on the textile industry, and on African cotton producers.

Nearly two decades have passed since the collapse of the regime in 1974, yet researchers are still only beginning to unravel its intricacies. I hope therefore that this study will contribute to the growing body of research on the nature of the regime and provide insight into the development of the Portuguese textile industry. As more research is completed, perhaps we can revise our interpretation of Portugal's era of authoritarian rule. Additionally, while this study addresses a particular relationship within the political and economic history of twentieth-century Portugal, its findings intend to give researchers a broader understanding of the politics of colonialism, the articulation of industrial interests, and the utility of authoritarian economic solutions.

1

The Creation of the *Estado Novo*, 1926–1936

Chapter 1 provides a context for the relationship between the regime, the textile industry, and colonial cotton interests by outlining the difficulties of the Republican period, explaining the coup of 1926, and documenting the economic and political changes which followed. In asserting its control over metropolitan and colonial interests and issues, the new regime moved cautiously. The decade following the overthrow of the discredited Republican government in 1926 was a period of political consolidation and economic stabilization. The regime sought to reduce the political instability and economic disruption that had helped bring it to power by enforcing a nationalistic political programme while repressing political parties, and it appeased economic élites in society while destroying organized labour. Consolidation and stabilization lasted at least until 1936, yet the creation of government regulatory agencies, an early obsession with controlling the balance of payments, and colonial intervention foreshadowed some of the lasting characteristics of the regime.

THE COUP OF 28 MAY 1926

When a military coup on 28 May 1926 removed the beleaguered Republican regime from power, few were convinced of the new government's longevity. Noting that it was the fortieth Cabinet in fifteen years, the *Economist* of Great Britain remarked, 'the present change is not likely to be the last.'[1] While many in Portugal welcomed the coup, sixteen years of Republican rule had bitterly divided Portuguese society.

In the aftermath of the military takeover, these divisions were reflected in political conflicts over the characteristics of the new

[1] *Economist*, 5 June 1926, 1087.

regime. Four factions emerged. Monarchists appealed for the return of the Portuguese king who had been ousted by the proclamation of the Republican government in 1910. Political groups such as the Integralists (formed 1914) and semi-political study groups such as Centro Académico da Democraçia Christã (CADC—founded 1901, revived 1912) influenced by political ideas and experiments from France (particularly Action Française and the ideas of Maurras), Germany, and Italy, argued for the creation of a corporate governmental system.[2] Additionally, many Catholics, previously alienated by the Republican regime's staunch anticlericalism and its tacit approval of attacks against religious institutions and the clergy, called for a new understanding, even reunification of Church and State. Finally, the army, which had already demonstrated its interest in Portuguese politics by undertaking the coup, intended to remain actively involved in Portugal's political future.

Although the coup had been well-planned, once in office the military plotters who had overthrown the Republican regime had no clear-cut strategy to ameliorate these divisions. Instead they opted to balance the political factions and to resolve differences through compromises. These early political compromises depict the conciliatory side of the regime in the decade after 1926, and they were incorporated into the 1933 Constitution.

To heal the bitter divisions, the new government advocated a nationalist approach to the country's problems. Constant changes of government and rotation of ministers had characterized Republican rule. Between 1910 and 1926 there were nine presidents, forty-five governments, and two dictatorships. It was not simply initial lack of experience, but also the lack of a coherent parliamentary majority and so the ceaseless struggle for power between Parliament and the Cabinet which was responsible for the political instability. According to a historian of the Republican period, A. H. de Oliveira Marques, 'Government entirely depended upon Parliamentary majorities: they often fell because of petty questions, personal dislikes, or whimsical actions. About twenty cabinets had to resign because of parliamentary distrust.'[3]

[2] Herminio Martins, 'Portugal', in S. J. Woolf (ed.), *European Fascism* (London: Weidenfeld and Nicolson, 1968), 302–5; and A. H. de Oliveira Marques, *History of Portugal*, ii (New York: Columbia Univ. Press, 1972), 157.

[3] Ibid. ii. 163.

Alternatively, by June 1930 the new government had created the National Union, a party which concentrated on unity rather than division, admitting any individual or group interested in collaborating with the regime and suppressing those who did not.[4] The Union was to replace the factionalism of the Republic and its programme avoided the issues that divided potential supporters of the new regime by calling for a national solution to the country's problems rather than one characterized by Republican or Monarchist, religious or non-religious solutions. When a group of Republicans and Socialists sought guarantees of free and competitive elections to government posts, those now in power responded that opposition was possible as long as it did not attack the government. Also, since organized political parties had proved so destructive in the past, opposition must take the form of 'currents of opinion' instead.[5] In effect, the party system was dead. One national party was to replace the several parties that had gone before.

The National Union (União Nacional) was designed to collaborate with and support the regime and not to participate in or lead it as single parties in other countries were to do. But then, it had not helped to seize power; rather it was an after-thought designed to consolidate power after a successful coup. Only 50 per cent of the deputies in the first National Assembly and the Corporative Chamber in 1935 were members of the National Union.[6] The National Union had a limited role in organizing and transmitting ideas on behalf of the government, but by and large it was not meant to mobilize mass support nor was it designed to become an all-powerful totalitarian party. Its primary goal was to foster nationalism—a major building-block of the new regime. Government policy was, by definition, national; opposition to it anti-national, and therefore, illegitimate.

Key elements of the new nationalism were Portugal's Asian and African colonies. Although the Republicans had promised to restore Portugal's imperial stature, in fact, they had badly neglected the colonies. In Portugal's two most important colonies of Angola

[4] Manuel Braga da Cruz, *Monárquicos e republicanos no Estado Novo* (Lisbon: Publicações Dom Quixote, 1986), 89.

[5] Ibid. 96.

[6] Ibid. 100.

and Mozambique, promises of greater autonomy and decentralization had only resulted in financial disaster and settler outrage—two issues which received constant media attention during the 1920s. Angola had a debt of £4,500,000 and unauthorized printing of notes by the Banco Nacional Ultramarino (BNU) had led to inflation and scandal.[7] Mozambique was a law unto itself, run by foreign-owned concessionary companies with their own economic and political agendas. Additionally, exports to the colonies, on which both Portuguese agricultural and industrial exporters depended, had actually declined from 19 per cent of total exports in 1910 to 14 per cent by 1920.[8]

Far more threatening than financial losses arising from debt and decreased trade was the publication of the Ross Report in 1925.[9] The Report contained extensive details of Portuguese colonial labour abuses which renewed talk amongst the imperial powers of a possible takeover of the Portuguese colonies to 'save' them from Portugal. The loss of Portugal's overseas possessions not only alarmed commercial interests worried about the decline in colonial trade, but also incensed and humiliated the Portuguese élite and the military. Having an empire was a part of Portugal's heritage, losing it would be an insult to national pride and would damage the concept of the Portuguese nation. For the military, the possession of the colonies and their own existence were symbiotic: military history, its prestige, and career prospects derived from nineteenth-century colonial conquests; giving them up was unthinkable.[10]

The nationalist solution proposed by the military and the *Estado Novo* which followed it sought to restore the colonies to their rightful place in Portuguese national identity. Colonial rhetoric after 1926 emphasized Portugal's imperial greatness. A prominent theme in the *Estado Novo*'s ideology stressed the indivisibility of the Portuguese metropole and its colonies. As Salazar indicated in his speech to the National Union in 1935, a long history of colonial possessions distinguished Portugal from other nations, contributed

[7] Douglas L. Wheeler, *Republican Portugal: A Political History 1910–1926* (Madison: Univ. of Wisconsin Press, 1978), 217; and Clarence-Smith, *The third Portuguese empire*, 128.

[8] Armando Castro, *A economia portuguesa do seculo XX (1900–1925)* (Lisbon: Ediçöes 70, 1973), 285.

[9] Wheeler, *Republican Portugal*, 218.

[10] Ibid. 188.

to the country's independence, and was a major component of its heritage.[11]

Alongside its nationalistic appeals, the government agreed to a second compromise: a controlled presidential election. An election would appease the various factions within and without the government and provide a badly needed veneer of legitimacy.[12] Having a presidential election would indicate the government's respect for some Republican practices. On the other hand, the new President would be elected by direct universal male suffrage (of which the short-lived regime of Sidónio Pais in 1917–18 had provided a precedent), rather than by Parliament as under the Republic.[13] This satisfied the Monarchist demand that Presidents be free from governmental responsibility and not dependent upon the legislature. Like a monarch, the President was perceived to be above Parliament.[14]

On 25 March 1928 General Oscar Carmona was elected to the presidency, thus confirming the nationalistic and Caesarian character of the new regime and its willingness to compromise amongst the different factions within Portuguese society. At this time, a third compromise between Church and State was arranged. Aware of a still strong current of anticlericalism, the military sought to maintain the legal separation of Church and State accomplished by the Republicans. Whereas Republicans had alienated the Catholic Church with their anticlericalism, the military sought its co-operation, by no longer pursuing an aggressively anticlerical policy. Moreover, the new government sought the personal collaboration of political Catholics, most notably by appointing Antonio Oliveira Salazar to the Ministry of Finance in 1928. Salazar was an expert in finance at Coimbra University, and a prestigious member of CADC and also Centro Católico, a conservative Catholic pressure group formed in 1917 to promote the ideas of Social Catholicism, to debate the country's difficulties, and to advocate greater cooperation between Church and State. Composed of intellectuals, Centro Católico adhered to the Social Catholic principles of social

[11] Antonio de Oliveira Salazar, 'Discurso', *I Congresso da União Nacional*, Lisbon, 26–8 May 1934, i (Lisbon: Edição Nacional, 1935), 68–9.

[12] Richard Robinson, *Contemporary Portugal: A History* (London: George Allen and Unwin, 1979), 43.

[13] Wheeler, *Republican Portugal*, 139–46. Voting rights were later restricted to heads of families, see Robinson, *Contemporary Portugal*, 48.

[14] Cruz, *Monárquicos e republicanos*, 91.

unity, the political principles of corporatism, and the values of the nation, the family, authority, and hierarchy—ideas which were becoming increasingly popular in Portugal and in other European countries.[15] Salazar's appointment to the Ministry of Finance in 1928 and his subsequent nomination as Prime Minister in 1932 were therefore highly significant. Not only did they indicate that the regime intended to associate closely with the Catholic Church, but also that the regime endorsed the principles of Social Catholicism and corporatism. Most importantly, since Salazar was noted for his ability in financial matters, his early rise to prominence was an admission that the military could not handle its greatest challenge—the Portuguese economy.

ECONOMIC INTERESTS AND CONSTRAINTS

Economic issues and the divisions amongst economic élites were at least as responsible as political instability for the downfall of the Republican regime. By 1926 pressures brought by conflicts among agricultural, industrial, and commercial interests, coupled with general decreases in trade, increases in the cost of living due to inflation, and the government's deplorable financial position created demands which the Republicans could not satisfy, either politically or economically. Republican rule was attacked by landowners, financiers, industrialists, traders, and workers, most of whom welcomed the military takeover. The discussion below gives the origin of these difficulties in order to explain the purpose of Salazar's policies as Minister of Finance and later Prime Minister, and to provide a context for his later actions with regard to colonial cotton and the Portuguese textile industry. In turn, Salazar's initial acts defined some of the enduring priorities of the regime.

The Republican period brought Portugal to the brink of financial ruin. While the initial years of Republican rule were relatively stable and the budget was balanced from 1912 to 1914, Portugal's entry into World War I precipitated acute fiscal difficulties. Portugal's budget was in deficit from 1914 and was not balanced until 1923–4.[16] Much of the government's finance was devoted

[15] Martins in Woolf (ed.), *European Fascism*, 305; and Marques, *History of Portugal*, ii. 157.

[16] A. H. de Oliveira Marques, *A Primeira República portuguesa* (Lisbon: Livros Horizonte, 1971; 2nd edn. 1975), 131.

to the war effort leaving little expenditure for social services or economic development either at home or in the colonies. Total war costs were estimated at between £60 and £80 million, £10 million of it in Africa—a huge sum for Portugal, most of which was raised through borrowing or printing money.[17]

Following the war, the government resorted to credit from the Bank of Portugal to finance a consumer spending spree and to create bureaucratic jobs for Republican patrons. Added to the war debt then, the Republican government had an internal debt of approximately 300,000 contos which it owed to the Bank of Portugal.[18]

This led to galloping inflation and a rapid devaluation of the escudo. Between 1914 and 1917, prices rose 62 per cent and between 1914 and 1920, 452 per cent (Fig. A1).[19] Unable to raise taxes substantially, the Republicans could only print more money. Whereas in 1914 there were 5.66 escudos to the pound; by 1924, there were 133.95 escudos to the pound (Table A1). In the colonies inflation also worsened as Portuguese colonial commissioners borrowed heavily from the BNU, the bank of issue for the colonies.[20]

Ironically, devaluation rendered Portuguese exports more competitive abroad, and domestically, acted as a kind of protectionist tariff. It allowed Portuguese agriculture and industry to replace with domestic products those goods which had become too expensive to import. But domestic substitution of some products and increases in exports were not sufficient to compensate for the parallel increase in the cost of imports which could not be replaced, and the rise in machinery and raw materials purchased to feed industrial growth. Between 1920 and 1925, the Republican government therefore had a trade deficit of 75,000 gold contos, up from an average deficit of 30,000 gold contos for the period from 1905 to 1915.[21]

As the Monarchy had done before it, the Republican government offset the deficit in the balance of trade with re-exports of colonial

[17] Clarence-Smith, *The third Portuguese empire*, 117.

[18] Antonio José Telo, *Decadencia e queda da I República portuguesa* (Lisbon: A Regra do Jogo, 1980), 15.

[19] Castro, *A economia*, 61.

[20] Clarence-Smith, *The third Portuguese empire*, 118.

[21] Castro, *A economia*, 146. Due to the rapid devaluation of the escudo, gold escudos or contos (1,000 escudos) were often used to indicate real prices, since they referred to a fixed rate of exchange between the escudo and the price of gold.

produce via Lisbon and remittances from Portuguese citizens who had emigrated to Brazil. But since 1913, the re-export trade had steadily declined leaving emigrants' remittances as the main source of finance to shore up the deficit. Emigration, both clandestine and legal, had progressively increased throughout the 1800s as Portuguese peasants repudiated an existence of poverty and responded to promises of a better life in Portugal's former colony of Brazil. Portuguese governments turned a blind eye to the steady trickle of emigrants because they were able to use remittances sent back to family and relatives to cover chronic trade deficits. By the time the Republicans came to power, the trickle of persons to Brazil had become a flood. Having dropped off considerably during the war, in the post-war period the numbers of emigrants returned to the extraordinarily high levels attained in the first four years of Republican rule. Official figures (not including clandestine emigration) for this period show that between 1910 and 1914, 291,480 people emigrated and between 1919 and 1926, 301,145 people emigrated from Portugal.[22] Estimates indicate that remittances contributed approximately 16,000 to 18,000 gold contos to the balance of payments, but even this was not enough to cover severe trade deficits following the war.[23] Moreover, the emigration of entire families from 1920, rather than male heads of household as previously, probably acted to reduce the value of remittances sent back and further undercut government attempts to reduce the deficit after the war.[24]

The situation initially worsened as Salazar took over at the Ministry of Finance. Salazar's early struggles with the balance of payments became a lasting obsession throughout his lifetime and one with profound impact on the cotton policy, but at this time he was unable to exercise much control. Brazil's decision to withdraw financial support of its immigrant recruitment agencies led to a drop in emigration from Portugal and therefore, of remittances to cover the payments deficit. By 1931, in the midst of the Depression, Brazilian authorities restricted the sending of remittances and prohibited emigration from Portugal altogether. The effects were dramatic. From 23,196 emigrants in 1930, emigration dropped to

[22] Joel Serrão, *A emigração portuguesa* (Lisbon: Livros Horizonte, 1977), 31–2.
[23] Castro, *A economia*, 146; and Miriam Halpern Pereira, *A política portuguesa da emigração, 1850–1930* (Lisbon: A Regra do Jogo, 1981), 37.
[24] Halpern Pereira, *A política*, 56.

6,033 people in 1931 and did not begin to increase significantly until after World War II.

The loss of emigrants' remittances to cover the balance of payments deficit then made it imperative that the new government seek alternative solutions to the payments deficit. Yet in formulating economic policy, Salazar had to arbitrate carefully amongst the conflicting demands of economic élites based in agriculture, industry, and commerce, whose differences were as great as those among the political groups and who had successfully thwarted Republican attempts to stabilize the escudo or to alter the financial structure of the country—in the end—by abandoning the regime.

Agricultural difficulties and interests evoked particular concern. Agriculture still accounted for 50 per cent of Portugal's exports and occupied 50 per cent of Portugal's workforce in the 1920s.[25] After the war, a number of problems plagued agriculture which the Republicans could not correct and which now demanded the new government's attention. First of all, inflation led to a drop in domestic food consumption of some goods, and demobilization reduced agricultural exports of others (notably wine and meat) as they were diverted to returning troops. Secondly, after strikes and near starvation in the cities due to a wheat shortage, the Republicans abandoned their protectionist policies for Portuguese-grown wheat in 1920 and resorted to imports, undermining sales of more expensive Portuguese wheat, and earning the enmity of large landowners in the Alentejo.[26] Though wine exports subsequently recovered and the government attempted to pacify landowners by protecting Portuguese wheat again in 1923, inflation continued, forcing up salaries and reducing domestic consumption. Additionally, though much discussion had taken place on the subject, no agrarian reform had occurred to improve the meagre return that agricultural workers and peasants received for their produce and/or their labour. As a result, many agricultural workers and peasants, particularly from the North, swelled the numbers of those who emigrated to the cities or to Brazil.

The drop in exports, inflation, chops and changes in the policy of protectionism for Portuguese wheat, constant debates about agrarian reform, and the resurgence of emigration affected both

[25] Telo, *Decadencia*, 37.
[26] Marques, *A Primeira República portuguesa*, 17–21.

those large, Southern landowners (about 50 per cent) whose major source of income depended on rents from sharecroppers who farmed their land, and commercial enterprises in the North and South which depended on cheap agricultural labour and exports for their survival.[27] Refusing to change traditional patterns of land ownership and land use, landowners had sought to block all Republican attempts to alleviate poor living standards in agricultural areas and to improve productivity.[28]

Simultaneously, the ability of agricultural interests to dominate the political and economic agendas of successive governments had been eroded by the growth of industry in both traditional and modern sectors. Although the distinction is not a scientific one, nevertheless, a country's level of industrial development can be gauged by the kind of industry it has, either 'traditional' or 'non-traditional' (modern). Traditional industries are labour-intensive industries and are generally the first to develop in a nation which is industrializing. These include textiles, leather, foodstuffs, beverages, tobacco, resin, cork, soap, and cutlery. Non-traditional or modern industries tend to be capital-intensive and include fertilizers, electrical equipment, oil-refining, cement, paper and paper products, chemicals, metal-working, transport equipment, and possibly plastics.[29] Some or all of these modern industries are founded after a nation has an educated workforce, sufficient capital for investment, appropriate technology, and an adequate infrastructure.

After 1870 Portuguese industry showed consistent growth, but it posed little threat to dominant agricultural interests. With the onset of World War I, however, both industrial production and trade rose. Increases in exports to the colonies and abroad brought large profits for the traditional industries of wine, textiles, cork, fish, and conserves. Simultaneously, several modern industries such as fertilizers, cement, and electrical materials had been founded to replace shortages experienced during the war (Table A3).[30] These

[27] Fernando Medeiros, *A sociedade e a economia portuguesas nas origens do Salazarismo* (Lisbon: A Regra do Jogo, 1978), 130; and Marques, *A Primeira República portuguesa*, 23–5.

[28] Medeiros, *A sociedade*, 339.

[29] See pp. 172–3 in V. Xavier Pintado, *Structure and Growth of the Portuguese Economy* (Geneva: EFTA, 1964).

[30] Rosas, *Nos anos trinta*, 58.

industries not only introduced a series of new demands on the government but more importantly, their growth fragmented Portuguese society, shattering the Republic's usual base of support.

Examining first the demands by industry, these ranged from a request for protected markets to threats of retaliation against organized labour. When the Republican government did not satisfactorily meet any of these demands, industrialists withdrew their support. The split betwen government and industry began shortly after the war. Following the short boom of 1919–20, economic conditions worsened in most of Europe and governments responded with protectionism. Portugal's foreign and colonial markets contracted or were re-entered by foreign competitors. Traditional industries, many of which had expanded during the war, found themselves with excess capacity and few markets, not even in the colonies, in which to sell their production. Although remittances from the steady stream of emigrants who had left Portugal since the beginning of Republican rule actually led to increased internal demand for some consumer goods, Portuguese industry did not yet possess the technical capacity to satisfy it and demand was thus met by resorting to imports.[31]

The Republican government offered little support to industry. It was both unwilling and incapable of either protecting Portuguese industry in the domestic market or retaining foreign and colonial markets. Manufacturers, who previously had benefited from Republican aid during times of crisis, felt betrayed by the government and began to oppose it.

Subsequently, even when the currency was stabilized and the deficit reduced in 1923–4, industrialists still criticized government policy through their associations and a newly formed conservative pressure group, the União dos Interesses Económicos (Union of Economic Interests—UIE).[32] Though devaluation had increased the cost of imports, simultaneously it had given those industries who exported a slight edge in foreign markets. During the period of rapid devaluation after the war, it was customary for manufacturers to purchase their imports early and sell their exports late in order to benefit from any devaluation which had occurred between the beginning and end of production. When the escudo stabilized in 1924, Portuguese industrialists, notably in wine and

[31] Medeiros, *A sociedade*, 351.
[32] Wheeler, *Republican Portugal*, 210.

textiles, lost this advantage, lost the few foreign markets in which they traded, and reverted to selling their products domestically, resulting in cutthroat competition in the internal market.[33]

Simultaneously, the inflation, debt, and bank scandals enveloping the colonies reverberated in the metropole. By 1924 the colonies were unable to pay for goods they had imported from Portugal, paralysing financial transfers, and sending many businesses, particularly those in the textile industry, into bankruptcy. Had circumstances been normal, most industries would have resorted to credit to cover the colonial transfers but many banks, which had arisen to speculate on foreign-exchange markets during times of devaluation, had now shut or were experiencing severe capital shortages due to government controls and attempts to stabilize the escudo.[34] The Republicans tried to offset the damages caused by stabilization and the paralysis in colonial transfers. They reserved the domestic market for Portuguese industry with a requirement that all customs duties on imports into Portugal be paid in gold. Unfortunately, they had already lost the loyalty of many industrialists.[35]

Work stoppages occasioned by the growth of organized labour moreover turned industry actively against the government. Social agitation after 1919 included bombings, sabotage, and murder, and terrified the urban bourgeoisie, bureaucrats, landowners, and industrialists. Between 1919 and 1925 there were 148 strikes, many of which erupted into violent clashes. In response, employers formed their own organization, the Confederação Patronal (employers' confederation) followed by the União dos Interesses Económicos (UIE) in 1925.[36]

Of great concern to the Republicans, and later to Salazar, was that industrialization had not only produced a new set of demands, but also had irrevocably shifted the balance of power within the Portuguese élite. Hand in hand with the industrial growth during the war came the emergence of powerful families who dominated certain sectors and the formation of embryonic links between

[33] Telo, *Decadencia*, 290.

[34] Ibid. 285.

[35] Ibid. 231.

[36] Marques, *History of Portugal*, ii. 138–9; and see Rogério Roque Amaro, 'O Salazarismo na lógica do capitalismo em Portugal', *Análise Social*, 18: 72–4 (1982–3), 1003, on strikes between 1910 and 1925 and the role of the working class under the Republic.

finance capital and industry. The Sommer family in cement, Narciso Ferreira in textiles, and Alfredo da Silva in superphosphates and shipbuilding exemplified the concentration which was occurring in some industries.

In addition, links were formed between banking capital and industry, particularly in the fields of fertilizers, flour-milling, and metallurgy.[37] A short boom in 1919–20 reinforced these trends: new factories were built and eleven new banks, associated with import/export businesses and commercial houses, were founded at this time.[38] The post-war crisis resurfaced in 1922 leading to the closure of fourteen banks (six of which had just opened during the boom), which further contributed to economic concentration. By this time, several large groups with a variety of commercial, industrial, and financial interests were becoming visible in both the domestic and colonial economies. Companhia União Fabril (CUF) under the direction of the politically powerful Alfredo da Silva, had agricultural interests in the colonies, links with banks and import/export businesses, and a monopoly on chemical fertilizers. The Burnay family had its own bank, the Banco Burnay, an interest in the Companhia dos Tabacos—the Portuguese tobacco monopoly; mineral exploration in Angola including the large diamond concession, Diamang (also financed by Belgian capital); and agricultural and commercial interests in Angola and Portuguese Guinea. In 1925 the Banco Burnay merged with the Banco Fonseca, Santos e Viana (which also had interests in colonial companies), and the Lisbon and Azores bank. The Espírito Santo family also had its own banks, the Banco Espírito Santo and the Banco da Beira, an interest in Diamang, and agricultural companies in Mozambique and Angola. Less powerful but with important regional interests, Borges e Irmão was a Northern banking house founded in 1884 by António Nunes Borges and his brother in the city of Oporto. Considered the bank of the North, its interests were in wine, textiles, and metallurgy. Banco Sotto Mayor, another large group, became the Banco Pinto e Sotto Mayor in 1925 with interests in Brazilian commercial firms, colonial commerce, particularly with Mozambique, and English coal mines. It also held a majority of the shares in a Portuguese electrical company.[39]

[37] Marques, *A Primeira República portuguesa*, 34.
[38] Ibid. 129–30.
[39] Medeiros, *A sociedade*, 98; and Rosas, *Nos anos trinta*, 85–6.

The emergence of powerful families within the industrial sector and the formation of embryonic links between finance and industry, and colonial trade and industry, brought political pressures to bear upon the regime. The growth of industrial and financial interests challenged the political hegemony of agriculture and small commerce and produced conflicts between and amongst the various economic sectors.[40] These conflicts fragmented the political power of the Portuguese élite. For example, in the grain-processing industry, industrialists preferred to import cheaper wheat from abroad rather than to buy it from Alentejo landowners.[41] The decision by the Republicans to protect domestic wheat producers in 1923 angered millers and divided their interests from those of agriculture. Moreover, along with industrialization came wage work, mechanization, and urbanization to which agricultural interests were radically hostile.[42] Labour was flocking to the cities, especially the Lisbon–Setúbal area.[43] Furthermore, increased trade with the colonies during the war and the immediate post-war period (especially in primary products such as the import of colonial wheat and rice) competed with domestic agriculture and introduced conflict between agricultural interests and traders. Finally, the desires of importers and exporters for free trade directly opposed those interests who lobbied for protected markets, both in industry and in agriculture.[44]

Unable to address the demands or reduce the conflict among élites, the Republicans were deserted by them and by their own left-wing, who were equally dissatisfied with the slow pace of reform. The Republic succumbed to a military coup in 1926. Economic policies adopted after the coup indicate that the new regime was not only willing to arbitrate amongst these competing interests, but also ready to intervene to effect financial reorganization of the economy. Faced with bankruptcy, the military government initially sought a loan from the League of Nations, but the unacceptable conditions attached to its approval 'functioned as an

[40] Medeiros, *A sociedade*, 71; and for a different interpretation, see Kathleen C. Schwartzman, *The Social Origins of Democratic Collapse* (Lawrence: Univ. Press of Kansas, 1989), chs. 5 and 6.

[41] Rosas, *Nos anos trinta*, 120.

[42] Medeiros, *A sociedade*, 302–4.

[43] Ibid. 114.

[44] Ibid. 113.

indirect external guarantee for the application of Salazar's orthodox and deflationary financial plan'.[45]

Salazar meant to reorganize Portugal's finances and reduce the rate of change in Portuguese society. Even before the Depression, balanced budgets, a stable currency, protectionism, and exchange and trade control became the hallmarks of government policy, as they would become for other countries during the Depression. Salazar introduced successive increases in customs duties from 1928 in order to protect the domestic and colonial markets, and stabilized the currency by fixing it in relation to the British pound.[46] Although he was unable to reverse the negative commercial balance until World War II, he did succeed in balancing the budget.[47]

It is probable that Salazar, like the Republicans, would have met resistance to protectionism and stabilization policies from export-oriented industry and agriculture, as well as commercial interests, had the Depression not intervened, belatedly, in 1931–2; instead it merely worsened an already critical internal situation. In addition to a drop in the number of emigrants and their remittances, the value of raw materials plummeted, seriously affecting colonial exports such as rubber, sugar, and coffee.[48] Equally, the value of Portuguese exports such as fish conserves, textiles, cork, resin, and wine dropped steeply with shocking effects on the balance of trade and on producers' profits.

Although on the whole the Portuguese economy suffered less than those of other European countries,[49] the negative effects of the Depression nevertheless allowed Salazar to further his autarkic and interventionist aims and to exercise strict control over trade. In a 1930 speech, Salazar justified strong state participation as necessary to order and co-ordinate the economy and develop production.[50]

[45] Martins in Woolf (ed.), *European Fascism*, 314; and Rosas, *Nos anos trinta*, 66.

[46] Ibid. 125; and Daniel Bessa and Carlos Pimenta, 'A economia sob o Estado Novo', in Antonio Costa Pinto *et al.*, *O Estado Novo: Das origens ao fim da autarcia, 1926–1959* (Lisbon: Editorial Fragmentos, 1987), i. 302. Salazar even tried to return to the gold standard but had to abandon it when Britain went off the gold standard in 1931.

[47] Rosas, *Nos anos trinta*, 99. But see Robinson, *Contemporary Portugal*, 43. The League of Nations found the method of balancing the budget very dubious because pluri-annual items of expenditure were considered extra-ordinary rather than ordinary expenditure.

[48] Rosas, *Nos anos trinta*, 101.

[49] Ibid. 103.

[50] Salazar, *Discursos*, i. 74.

Intervention was used not only to increase the power of the state, but to secure support for the regime by protecting and aiding economic groups. The government intervened more directly in the economy to help both agriculture and industry cope with the crisis. For agriculture, an ill-considered wheat campaign was launched to benefit Southern landowners, whose own production had been intermittently undermined by cheaper imports, and to improve the balance of payments which had been negatively affected by the cost of wheat imports.[51] On the initiative of the state, an organization of wheat producers, the Federação Nacional dos Produtores de Trigo (FNPT) was created in 1932 to co-ordinate wheat production, set prices, supply credit to farmers, and give technical advice.[52] In addition, the Consorcio Português de Sardinhas and the Casa do Douro were formed in 1932 'to guarantee minimum prices of sale, facilitate credit and discipline competition' for sardines and Douro wine, respectively.[53] These two products figured importantly in exports, accounting each for 21 per cent of total exports and had been affected by the crisis. Not surprisingly, the government sought improvements in these sectors. For other kinds of wine and fruit, similar organizations with similar functions were also created. These organizations were precursors of the corporative organisms which would later intrude upon almost every sector of the economy.

Accompanying the creation of organizations for certain affected areas, the government passed legislation to protect and control industry in order to reduce internal competition which had grown since World War I and had been exacerbated by the Depression. Initially, the legislation was limited to a few selected, high-profile industries such as cereal-processing, resin, and conserves which encountered fierce domestic competition from foreign products, were linked to agriculture, or figured in exports. The reasons for this focus were numerous. In the case of cereal-processing, the regime intervened in order to reduce the competition inside the industry, but also to shift the industry from wheat imports to the purchase of Portuguese-grown wheat. In metallurgy, the intervention was effected to reduce imports of expensive machinery.

[51] Rosas, *Nos anos trinta*, 160–6.
[52] Manuel de Lucena, 'Sobre a evolução dos organismos de coordenação económica ligados à lavoura', Part I, *Análise Social*, 14: 56 (1978-4), 842.
[53] Rosas, *Nos anos trinta*, 98.

In fertilizers, it was to guarantee CUF the dominance of the industry.[54]

In 1931 the regulations were extended to all Portuguese industries. Facing precipitous drops in prices and limited market opportunities, many industrialists in Portugal—as elsewhere in Europe—welcomed government regulation at this time.[55] Passed as a temporary measure, the Industrial Regulation Law (Condicionamento Industrial), Decree-Law 19,354 of 3 January 1931 became a permanent feature of the regime throughout its existence. The law intended to avoid:

> the foreign control of industries, the excess or lack of competition, the deleterious consequences of poor geographical distribution, failures due to the lack of a technical base or financial guarantees, and the grave uncertainties to which the working class is exposed when its existence is linked to the precarious life of poorly equipped factories.[56]

The government proposed to carry out this regulation of industry by requiring that authorization be granted by a government department, the Direcção Geral da Indústria (DGI) within the Ministry of Agriculture, Commerce and Industry for the substitution or transference of machinery, and the expansion or creation of new factories. Subsequent decrees in 1931 and 1932 expanded the number of industries subjected to Industrial Regulation.[57] We shall examine the initial effects of this legislation on the textile industry in the next chapter.

In conjunction with its efforts to stabilize the Portuguese economy and reduce conflict between sectors, the government established more centralized metropolitan control over the colonies. It advocated stronger trade links between the metropole and the colonies to benefit commercial interests in Portugal and to satisfy its own balance of payments priorities. After the Stock Market crash of 1929, Salazar also assumed the post of Colonial Minister and

[54] Ibid. 181–3.

[55] Ibid. 152.

[56] José Brandão de Brito, 'Corporativismo e industrialização: Elementos para o estudo do Condicionamento Industrial', *Ler Historia*, no. 6 (1985), 54.

[57] Decree-Laws 19,409 of 4 Mar. 1931 and 20,521 of 17 Nov. 1931, from Rosas, *Nos anos trinta*, 291. The Ministry of Agriculture and Commerce and Industry was changed to the Ministry of Agriculture and the Ministry of Commerce and Industry in 1933; in 1939 the Ministry of the Economy combined the two with a separate Directorate of Commerce and Industry and a Directorate of Agriculture.

vigorously pursued the interrelated goals of internal financial stability and colonial integration, though the creation of the cotton regime illustrates that centralization and integration evolved slowly. As Minister of the Colonies, Salazar himself drafted the Acto Colonial (Colonial Act)—the decree integrating the metropole and the colonies. The main objectives were the establishment of greater metropolitan control over colonial finances and administration, and greater protection for trade between the colonies and the metropole. In order to save on foreign exchange, the colonies were to supply Portugal with raw materials, and in return, the metropole would be guaranteed markets for its manufactures. In addition, exchange control was introduced in Angola and Mozambique in 1931–2, which put an end to the paralysis of transfers which had occurred previously. Foreign imports into the colonies and the exchange to cover them were strictly controlled, forcing colonial buyers to purchase from Portugal. These measures helped Portugal's own trade balance and also appeased those industries which had suffered a loss of markets as a result of post-war developments and/or the Depression. Salazar's goals were not unusual but they were innovative: most colonial powers were adopting autarkic approaches to surmount the difficulties of the Depression, but few advocated exchange control as early and for as long as the Portuguese.

From 1928 the regime successively raised duties on foreign goods imported into the Portuguese colonies and granted preferential tariffs to Portuguese exports in order to favour the purchase of metropolitan goods. In 1932 the value of Portuguese exports to the colonies returned to 1928 levels and increased thereafter. Additionally, the value of colonial imports into the metropole also increased from 1932 though slowly.[58] As we shall discover in the case of the Portuguese textile industry, integration took some time and did not always please all parties.

The colonies were supposed to serve the metropole but not to be a drain on it. Colonies were expected to balance their own budgets, and any loans had to be contracted through, and approved by, the metropole. On most financial matters, other than purely routine ones, the Colonial Minister had the ultimate say.

[58] Rosas, *Nos anos trinta*, 146–7.

THE PURPOSE OF SALAZAR'S ECONOMIC POLICIES

The objectives of the economic policies outlined above are fairly evident. One was to make Portugal more self-sufficient by establishing links with the colonies. Another was to restore the finances of the state and to stabilize the economy by reducing expenses and improving trade. Stability also served a political purpose. Protecting all groups, the regime was seeking legitimacy and support. This is illustrated by the fact that the government did not confine its protection to the wealthier classes, but also sought to appease labour by promising the continuation of rent control, stable (low) wages, employment, and rudimentary social-welfare programmes.[59] However, with the carrot of stable wages and employment, labour received the stick of systematic repression.

Amongst economic élites, the government adopted a different role. It sought to satisfy at least partly both industrial and agricultural, large and small interests, to a greater or lesser degree. The government addressed industrialists and landowners in policies and legislation. It made efforts to protect industry and to improve agricultural production. Various sectors of the Portuguese economy were weak and fragmented and welcomed or could not resist state intervention. Moreover, the fact that the state aided most groups suggests that no one group clearly dominated Portuguese society. As Fernando Rosas has written: 'The rural interests could not govern as they wished, but the industrial bourgeoisie could not yet govern as it desired.'[60] The state thus acted as a kind of arbiter in attempting to restore the national economy, but to contain change.

Maintaining the status quo did not imply the watchman state. The methods by which the state aided industry and agriculture, and appeased employers and workers suggest that it intended to intervene actively in Portuguese society. Using the Depression and its effects as an excuse to interfere, the government created bureaucratic organizations and legislative controls to effect massive intervention in the Portuguese economy. In spite of Salazar's later bows to corporatism, the regime had already legislated a prominent economic role for itself by the early 1930s. Police repression and

[59] Ibid. 122.
[60] Ibid. 150.

bureaucratic control of labour, legislative limitations on the construction and expansion of industry, and an active commitment to trade-deficit reduction show that the government was already weaving its way into the pattern of Portuguese life.

THE LEGAL CONSTRUCTION OF THE *ESTADO NOVO*

In effect, the antecedents for the *Estado Novo*, which was legally constituted in 1933, were hammered out in the struggles which took place after 1926. The method of direct election of the President, the inclusion of Monarchists and Republicans in the government, the understanding with the Church, the ideology of nationalism, the creation of the National Union, greater state intervention, and the appeasement of various élites during the initial years of the military dictatorship laid the foundations for the future state. Thus, the formation of the *Estado Novo* was neither preplanned nor inevitable:

> The construction of the *Estado Novo* was, from an ideological point of view, the result of various ideological currents, which were amalgamated in it. Coming out of a military coup, whose political intention was more the negation of a regime than the clear affirmation of an alternative, the *Estado Novo* ended up by combining various other political and ideological influences within an original Democratic-Christian matrix.[61]

The influence of the political compromises and economic policies implemented after the 1926 coup is evident in the Constitution of 1933. Rather than a static summary of principles and goals, the Constitution of 1933 was a hybrid of different ideological currents and political objectives of the government. Several tendencies influenced the orientation of the Constitution reflecting the economic needs and political requirements of the *Estado Novo*, Salazar's personal preferences, and the international context of the time.

The Constitution contained several vestiges of the earlier regime, possibly as a concession to its Republican constituency. Church and State remained separate (Part I, Section X), the direct election of the President was constitutionally acknowledged (Part II, Section

[61] Cruz, *Monárquicos e Republicanos*, 13.

II, Chapter 1), and neither the pretender to the Portuguese throne nor his relatives were allowed to run for office. Members of the National Assembly and local parish boards were also to be elected by direct (limited) suffrage. Civil liberties such as the freedom of assembly and association were also formally guaranteed (Part I, Section II, Art. 14).[62]

A distinctive feature of the Constitution of 1933 was its corporatist character. As elaborated in the Portuguese context, corporatism comprised the following beliefs. First, it emphasized the organic rather than the individualist nature of society. According to corporatists, society was not the sum of individuals, but a conjuncture of units, of which the most important was the family. A man's identity came from his position in a group rather than from his actions as an individual. It was to the group and the nation that he owed his loyalty. The notion of groups was extended to the workplace where factory workers would be organized into *sindicatos* (workers' associations) and agricultural workers or peasants into *casas do povo* (associations of rural workers and their employers), fishermen into *casas dos pescadores* (associations of fishermen), and landowners and industrialists into *grémios* (employers' guilds). These groups were seen as 'natural' in contra-distinction to 'unnatural' atomized individuals pitted against each other in a great class war. Salazar stated that what the new government was trying to do was to 'construct a social and corporative state in strict correspondence with the natural constitution of society'.[63]

A natural hierarchy supposedly governed society, and this hierarchy was to order political as well as social relations. The parties and factions of the Republic had tried to undermine this natural order; industrialization threatened to destroy it. The Constitution held that it was the nation as opposed to parties or factions which constituted political reality, and therefore, the national interest had to be made to prevail over particular interests. Authority came not from the people, but sprang from the natural order of a settled organic society, and was defined by the state. Part I, Section I, Article 6 declared:

[62] Portugal, Secretariado da Propaganda Nacional, *Political Constitution of the Portuguese Republic* (Lisbon: SPN, 1937) (hereinafter Portugal, SPN, *Political Constitution*).

[63] Salazar, *Discursos*, i. 87.

It is the duty of the state:

(1) To promote the unity of the nation and establish order according to law by determining and compelling respect for the rights and guarantees derived from morality, equity or law, for the benefit of individuals, families, local autonomous bodies and other corporate, public or private entities;

(2) To co-ordinate, instigate and direct all social activities and to promote due harmony of interests, subject to those of a private nature being lawfully subordinated to the general interest;

(3) To strive for improvement in the condition of the least favoured classes of society and to prevent their standard of life from falling below an adequate human minimum.[64]

Corporatist philosophy stressed continuity and history as essential to the preservation of stability. Tradition was seen as giving people a sense of belonging, a feeling that they were part of something that went beyond their particular interests and concerns. The tradition chosen and developed by the *Estado Novo* glorified Portugal's imperial past, a theme which was constantly reiterated throughout the life of the regime.

The political and ideological framework to be erected upon these beliefs was to encourage the preservation of the family and Christian values (as conceived by conservative Social Catholic theory), romanticize rural life and the association with *a terra*, specifically, the land of Portugal. Abstracting from the notion of the harmonious family, *grémios* and *sindicatos* would be created at work—the productive family; and *freguesias* or parish associations locally—the political family. At the national level, corporations were also envisaged to act as co-ordinating bodies between the government and industry, commerce, and agriculture.

The inclusion of corporatist elements in the Constitution sprang from the international trend towards corporatism and its advocates inside Portugal, Salazar's personal beliefs and his academic and theological training, and Portugal's history and traditions.[65] First, corporatism in Portugal was part of a broader European movement, from Fascist Italy to the successor regimes of central Europe to the intellectual debates in France and the political-religious movement of Social Catholicism. Theoretically, its emphasis on harmony,

[64] Portugal, SPN, *Political Constitution*.

[65] See Wiarda, *Corporatism and Development*; Robinson, *Contemporary Portugal*; and Franco Nogueira, *Salazar*, i (Coimbra: Atlantida Editora, 1977).

unity, and hierarchy offered solutions to the class conflict, individualism, and materialism believed to have been engendered by liberal capitalism and cited by Portuguese corporatists as having been the cause of the troubles in the Republican period. As other countries moved towards notionally corporate dictatorships, notably Portugal's former colony of Brazil as well as Germany and Italy, these trends influenced the political decisions adopted by the Portuguese regime in the early 1930s.[66]

Secondly, Salazar's background and academic preparation favoured the adoption of corporative doctrine. Prior to his matriculation at Coimbra, Salazar attended religious institutions for much of his life. Both Leo XIII's *Rerum Novarum* and Pius XI's *Quadragesimo Anno* (1931) were said to have been influential in the formation of his political philosophy.[67] Both works stressed the collaboration of labour and capital—a central tenet of corporatism—as well as the paternalistic role of the state in protecting the weak, a role analogous to that of the Church and its congregation. In addition, CADC and Centro Católico stressed the harmony and unity of classes in their analyses of Portuguese society.

Thirdly, the adoption of corporatism has been traced to Portuguese history. Wiarda refers to Portugal's 'historic, natural corporatist tradition'.[68] The influence of the doctrine of Social Catholicism and the historical existence of guilds—medieval precursors of *grémios* and *sindicatos*—have been cited also as evidence of a corporatist past.[69] Additionally, the short-lived plebiscitary regime of Sidónio Pais in 1918, with its corporatist approach, was said to be a direct precursor of the political framework elaborated in the Constitution of 1933.[70] The Pais administration certainly involved members of CADC and Centro Católico.

But if Salazar was attracted to the spiritual aspects of corporatism, equally he must have realized their practical use. Howard Wiarda

[66] Marques, *History of Portugal*, ii. 179.

[67] Wiarda, *Corporatism and Development*, 96–7; Robinson, *Contemporary Portugal*, 45.

[68] Wiarda, *Corporatism and Development*, 9.

[69] See *Indústria Têxtil*, 1: 1 (Nov. 1936), 5; and Wiarda, *Corporatism and Development*, 87.

[70] Martins in Woolf (ed.), *European Fascism*, 309; and Wheeler, *Republican Portugal*, 139–40. According to Wheeler, many members of Salazar's first Cabinet as well as the 1935 National Assembly and Corporative Chamber had served in the Sidónio Pais regime.

in *Corporatism and Development* claims that Salazar was a 'shrewd pragmatist'[71] and further argues that the stabilizing, self-regulating, self-sufficient elements of corporatism probably appealed to his penchant for balance and collaboration. The second claim is less convincing than the first. Far more was behind the creation of corporatism than the reasons cited above. Most revolutionary governments or successor regimes try to call upon 'traditions' in order to institutionalize and legitimize a new regime. The military and civilian leaders in 1926 and after were no exception. While it may be true that former Portuguese regimes and periods in Portuguese history displayed features characteristic of corporatism, most European countries could cite examples of corporatist behaviour or institutions. Additionally, Caetano himself, one of the architects of corporatism and later Prime Minister, admitted that Portugal's medieval guilds were weak.[72]

The Republican period alone suggested that Portugal was a very uncorporatist society, hence the zealous attempts by state propaganda to mask and alter its fractious nature under cover of the corporative ideology. By fostering the illusion of unity, family, and harmony, the regime intended to achieve its ends by rendering opposition not only illegal, but illegitimate and invisible. This seems to be the most convincing explanation for the adoption of corporatism in 1933, *after* the Depression had weakened economic actors, *after* the state had already defined its political objectives. *Estado Novo* propaganda should not be confused with a political culture, which, if anything, seemed more disposed to factionalism than unity, disagreement than co-operation. The *Estado Novo* was a political construct cobbled together at a moment when corporatist ideals were not simply fashionable, but seriously considered as answers to the political, social, and economic ills affecting all of Europe. The attraction of corporatism for members of the regime and their supporters was not only that it was able to draw on Portuguese traditions for legitimacy, and that it filled their Maurrassian prejudices; most importantly, it appeared to offer a solution for a sharply polarized society with severe economic problems.[73] How it subsequently developed both theoretically and

[71] Wiarda, *Corporatism and Development*, 80.

[72] Schmitter, *Corporatism and Public Policy*, 63 n. 8.

[73] João Ameal, *Construção do novo estado* (Oporto: Livraria Tavares Martins, 1938); J. Rodrigues de Mattos, *Corporativismo em Portugal* (Lisbon: Livraria Moraes, 1937) and *Condicionamento Industrial* (Lisbon: Livraria Moraes, 1940).

practically was the result of political compromises some of which we have already analysed, and requirements, even whims on the part of the regime and its participants.

Salazar's repeated utterances that he was creating a corporate state as well as the Constitutional statement that, 'The Portuguese state is a unitary and corporate republic' have led to a preoccupation with what kind of corporatism the *Estado Novo* tried to or did achieve. Abstracting from the Constitution, many analyses try to determine whether the Portuguese government under Salazar was truly corporatist or not.[74] These exercises are useful up to a point, but the political reality is more complicated. Even by the 1930s, the state was wearing several hats. It was pragmatic, interventionist, and nationalistic; many of its actions already did not conform to the corporatist model.

In addition, focusing on the corporatist aspects overlooks another major element of the Constitution, that which provided for extensive state intervention. Many of the corporatist tenets, such as the notion of hierarchy, can be used to justify strong state control but the Constitution itself provided amply for state intervention in other ways. For example, the state was quite clearly given a dominant role in the authorization of corporative organizations 'for intellectual, social and economic purposes' (Part I, Section IV, Art. 16), the protection of public opinion from 'distortion' (Part I, Section VI), the co-ordination of economic and social life, the direction of private economic operations 'in order to secure greater social benefits than would otherwise be obtained' (Part I, Section VIII, Art. 33), and the development of the national economic system (Part I, Section VIII, Art. 34). At the same time as he spoke of the value of group representation Salazar also noted that:

> like a great family or a great concern, the nation, for the protection of its common interests and the attainment of its collective aims requires a head to control it, a center of life and action.[75]

[74] See esp. the works by Manuel de Lucena, Howard J. Wiarda, and Philippe Schmitter cited in the notes to the Introd. Lucena and Wiarda seem overly concerned with Portugal's conformity to the corporatist model; Schmitter nearly dispenses with the model when confronted with the evidence, see my earlier comments in the Introd.

[75] Salazar quoted by Elisabeth Leeds in Thomas C. Bruneau, Victor M. P. Da Rosa, and Alex Macleod (eds.), *Portugal in Development: Emigration, Industrialization, the European Community* (Ottawa: Univ. of Ottawa Press, 1984), 17.

In other words, the family is subject to the dictates of an overbearing *paterfamilias*. Disguised in a corporatist language is an authoritarian state granted constitutional powers to repress independent labour organizations and create state-controlled ones, to intervene in private economic operations to promote and protect the 'common interest', to censor opinions and educational establishments, to break up opposition movements, and to impose wide-ranging fiscal and economic policies.

POST-CONSTITUTIONAL PRACTICE AND ECONOMIC POLICY

Up to 1936 the government took the initiative in establishing nominally corporative organizations, but equally it continued to play the role of arbiter between opposing economic groups. It was a hybrid system constructed in response to different political and economic pressures. Throughout the 1930s, the regime was institutionalized and legitimized by the creation of various governmental bodies and political and economic associations as well as by the enactment of various laws and decrees. At an institutional level, the Constitution provided for a President, but in fact the Prime Minister—Salazar—had most of the power; a National Assembly composed of 90 members (in 1934, only 6 per cent of the population was eligible to vote) and the Corporative Chamber, an amalgamation of the country's social, political, and economic interests consisting of over 100 members practically handpicked by Salazar.[76] Neither the National Assembly nor the Corporative Chamber had any effective power. A bureaucratic apparatus also was created to deal with the increased role of the government in political, social, and economic affairs and to provide patronage for supporters of the *Estado Novo*.

In its organization of Portuguese society, the government was most active in the area of labour. Decree-Law 23,050 established *sindicatos* or unions which industrial workers in each sector had to join: 191 existed by 1935.[77] Through the *sindicatos*, worker–

[76] Philippe Schmitter, 'The Regime d'exception' in Graham and Makler (eds.), *Contemporary Portugal*, 8. The Corporative Chamber did not legislate but only gave opinions (*pareceres*) on legislation, some of which we shall refer to later.

[77] Schmitter, *Corporatism and Public Policy*, 17.

employer relations were restructured and worker organization strictly controlled. In agriculture, *casas do povo* were created to promote the collaboration of landowners and peasants by including them in a single organization. In reality, landowners were often the directors and peasants had little power in the organizations. One hundred and forty-one *casas do povo* existed by 1935. Fishermen too were organized into *casas dos pescadores* (28 by 1937) designed to foster co-operation between employers and labour.[78]

The reorganization of labour and capital was accompanied by the Estatuto Trabalho Nacional (ETN), Decree-Law 23,048 of 23 September 1933, a comprehensive labour statute covering everything from strikes to salaries. The statute gave a very detailed description of the new relations between workers and employers and declared that 'the state had the right and obligation to regulate national social and economic life in the common interest'.[79] Through the ETN the government created the Sub-secretáriado de Estado das Corporações e Previdência Social (Undersecretariat of State for Corporations and Social Welfare) to address all aspects of labour such as labour contracts, social assistance, and social welfare. This statute became the basis of a plethora of legislation designed to keep salaries and unemployment low and labour protest to a minimum: in short, to pacify and control the workforce for the benefit of business and social stability.[80]

Employers and landowners were organized too, though to a lesser extent, and more slowly than labour, for the political priorities were different. Prior to the 1933 Constitution, the government had created organizations in several agricultural sectors. After 1933 several of these organizations were changed to fit in with the corporative institutions envisaged in the Constitution, but ostensibly their role remained the same, to organize producers, to oversee all aspects of the production process. For wine, conserves, fruit, wood, wool, wheat, and rice, membership of these organizations was compulsory according to Decree-Law 23,049 of 1933, but for most industries, membership of employer organizations (*grémios*) was voluntary after 1934, either because the government

[78] Ibid. 20 and 23.

[79] Wiarda, *Corporatism and Development*, 107.

[80] For a lengthier discussion of the law, see Lucena, *A evolução* i. 179–205; and also Portugal, Sub-secretáriado de Estado das Corporações e Previdência Social, *Problemas da organização corporativa*, no. 10, conferência realisada 13 Jan. 1934.

did not wish to reorganize that particular industry or in response to resistance from industrial groups.[81] As forthcoming chapters will show, this concession by the government marked the beginning of government favouritism to powerful industrial interests in society.

At least until 1936, the state continued to arbitrate amongst the various groups in economic matters. It invested in infrastructure, communications, and energy. In agriculture, price controls, protection, and advice were managed by the government-created corporative organizations. Domestic production of rice, wool, and wheat were given preference by the government over foreign imports though the wheat campaign still failed to satisfy demand by the late 1930s and was dropped in 1936.[82] High prices were set for national produce and imports were restricted if foreign goods were of a lower price. Rice, olive oil, cork, and resin all showed increased profits during the 1930s.

Though less attention was paid to industry than to agriculture, there was slow but steady progress. Customs regulations sought to protect established national industry from foreign competition; industrial regulation legislation attempted to prevent the construction of new factories. Medium and long-term credit for industrial and agricultural enterprises was made available through the Caixa Nacional de Crédito (founded in 1929).[83] Monopoly concentration was sustained in the case of tobacco and phosphates (CUF) or encouraged in many of the most important sectors of the economy such as cement, fertilizers, petroleum-refining, shipbuilding, and electricity. Sommer consolidated his control of the cement industry in 1934; in 1936 the shipbuilding monopoly passed to CUF.[84] The less concentrated industries—resin, textiles, glass, ceramics, and pharmaceuticals—were also protected.

The government was active in the colonies too, seeking to pacify colonial interests by tying them closer together with the Portuguese economy. In addition to making the colonies pay for themselves mentioned above, after 1933, metropolitan imports of colonial products such as sugar, coffee, cotton, and cacao were maintained

[81] Lucena, *A evolução*, i. 271.
[82] Rosas, *Nos anos trinta*, 165.
[83] Ibid. 190; and see figures on credit allocated between 1929 and 1936, 256.
[84] Rosas, *Nos anos trinta*, 214–17.

or increased, often as a result of legislation forbidding foreign imports of these products. Metropolitan exports to the colonies, particularly textiles and cement, also increased aided by similar legislation.[85]

Looking back at the period from 1926 to 1936, the new regime was trying to appease a variety of forces in Portuguese society: Republicans and nationalists, colonial traders and domestic industrialists, peasants and landowners. Caught economically and politically between highly articulate, but fragmented groups, the *Estado Novo* sought to establish its legitimacy by playing the role of arbiter, by offering something to nearly everyone. Particularly amongst the economic élites in Portuguese society, the *Estado Novo* was trying to balance interests. For peasants and workers, a carrot and stick approach operated.

The Constitution of 1933 reflected the contradictions embodied in these actions. The Constitution simultaneously praised private property yet enshrined state intervention. It advocated collaboration yet created separate organizations for industrial workers and employers. It maintained elective offices and formally guaranteed individual liberties but justified increased control; and sought integration between the metropole and the colonies.

The statist elements of the Constitution anticipated the time when the *Estado Novo* would intervene with impunity in every aspect of Portuguese society from the factory to the farmhouse, in the fixing of prices, the setting of norms, and the adoption of quotas. Following the passage of the Constitution the regime sought to legitimate and consolidate its economic and political position through organizational restructuring. By establishing institutional bodies and legislation designed to aid the gradual development of agriculture and industry, the *Estado Novo* intended not only to build economic and social support, but also to establish a presence in the economy in order to advance its own priorities of financial stability and national self-sufficiency. Economic change was slow, but institutional reorganization proceeded quickly. This was a period in which state power began to be defined and institutionalized; the next period would reveal how the power was used.

We turn now to an examination of the structure of the textile

[85] Ibid. 146.

industry, its background, historical development, and the difficulties it was experiencing at the time of the 1926 coup. We will discuss the initial attempts by the *Estado Novo* to appease textile industrialists while pursuing its own goals of financial equilibrium and social stability.

2

The Nature of the Portuguese Textile Industry

The military coup in 1926 symbolized a new era for the textile industry. Not only were its connections with the government altered but so also was its relationship with the colonies. After 1926, and especially after Salazar's rise to prominence, a triangular relationship among colonial interests, textile industrialists, and the government began to form. Like most such relationships it was extremely complex and often unbalanced, and took some time to be established.

In the period from 1926 to 1936 government interests centred around appeasing political groups in the country and stabilizing the economy. The government achieved some of these objectives by granting concessions, but also it restructured political expression, economic behaviour, and commercial opportunities. Many of these efforts coincided with the wishes of the textile industry at the time. For example, the industry wanted markets and the government wanted to improve the balance of payments: so an attempt was made—increasingly as state power was consolidated—to secure the colonial and domestic markets by eliminating imports of textiles and ginned cotton from abroad. As other countries were enacting similar measures in order to overcome the Depression, the government could justify its protectionism and interfere in markets, taxing foreign textiles and giving preferential tariffs for Portuguese goods.

Industrial and governmental desires additionally overlapped when it came to controlling growth in the industry to limit competition. Although Condicionamento Industrial was not initially successful, the creation during the 1930s of legal precedents and a rudimentary bureaucratic apparatus to administer it were important elements of future government intervention in the industry.

The ramifications of the laws, government intervention, and interests were another matter however. The effective application of avowed policies remained uncertain. On one level, they seemed to have little effect at least in the early 1930s. We shall discuss the price incentives for colonial cotton production in the next chapter, but it is evident that imports of colonial cotton into the metropole did not increase substantially until the late 1930s. Moreover, whilst textile factories were prevented from locating in the colonies, the government did not completely exclude foreign competition there until World War II. Finally, and most worrying of all, overproduction in the industry continued and the government seemed unable to control its growth and expansion.

At the level of power rather than policy, the implications of this new triangular relationship were more serious and more long term. Government intervention into the structure, markets, and prices of the industry had been established at a time when the interests of state and industry coincided and industry was weak. Unable to solve its problems on its own, the industry welcomed government aid. Once their interests began to diverge, the industry was no longer a free agent.

BACKGROUND

Portugal's position as a mercantile power in the sixteenth century profoundly affected its industrial structure and development. The early preference for colonial commerce and the profits and taxes to be made from it, encouraged government to be indifferent to the development of the mainland and relatively unresponsive to internal concerns. It also encouraged investment in commerce and overseas ventures, or in government paper, rather than in productive enterprises.

These factors, combined with a lack of raw materials, led to a neglect of the manufacturing sector and a reliance on Great Britain to supply and receive a majority of Portugal's imports and exports. Such dependence on Britain structurally affected Portuguese economic development in two ways. First of all, Portugal geared its exports towards items that Britain did not or could not produce. These tended to be mainly agricultural goods, minerals, and later, a small number of manufactured articles. In comparison with the volume and value of British goods imported into Portugal, the terms of trade for these exports were negative, which contributed to Portugal's chronic trade deficits. Secondly, because British imports were manufactured more cheaply than goods which could be produced domestically, they thwarted the growth of similar enterprises in Portugal.

Owing to protectionist legislation, technological improvements, successful lobbying by pressure groups, changes in government priorities, a renewed interest in the Portuguese colonies and other factors, Portugal embarked upon sustained industrialization during the latter part of the nineteenth century. However, growth rates between 1870 and 1913, which while respectable, were not sufficient for Portugal to reach the level of more advanced nations such as Germany and the United Kingdom. During this period, Portugal's annual growth rate was around 2.5–2.8 per cent whereas Japan, Russia, and Sweden were experiencing growth rates from 8 to 12 per cent between 1888 and 1913.[1] Portuguese industry enjoyed another spurt of growth during World War I but Portugal's place in the world economy remained unchanged. With regard to trade, as late as 1925, Portugal's major exports were port wine, dried fish, cork, and cotton cloth and Great Britain still accounted for 29 per cent of the combined value of Portugal's exports and imports.[2]

Before examining how the Portuguese textile industry emerged as a major industry within these constraints of the Portuguese economy, it is important to define what is meant by textile manufacture. In this study, textile manufacture refers only to the use of cotton and not to wool (a separate industry) nor synthetics (which saw a limited growth during the 1960s). Secondly, this study refers to the procedure whereby ginned cotton is converted to finished yarn and cloth, but not to the production of clothing which is considered a separate industry and which only really expanded in Portugal during the 1960s. Spinning and weaving are the two major processes involved in the conversion of ginned cotton to yarn and cloth; other processes include cleaning and combing raw cotton before it is spun and dyeing and printing cloth after it is woven.

In order to ascertain the size and growth of a textile industry over time, it is usual to refer to the number of spindles and looms. Because the processes of textile manufacture are usually inter-

[1] Jaime Reis, 'A industrialização num pais de desenvolvimento lento e tardio: Portugal, 1870–1913', *Análise Social*, 23: 96 (1987-2), 213–14.

[2] Rosas, *Nos anos trinta*, 68–9.

dependent and interrelated, an examination of the quantity and quality of the machines used for spinning and weaving generally gives a very good indication of the state of the rest of the industry as well as its structure. While this standard is applicable also to Portugal for most of the twentieth century, it insufficiently describes the original structure of the industry. Because of the importance of commerce to the Portuguese economy and the dominance of British goods in the domestic market, initially the textile industry was split vertically between a spinning and weaving branch which produced low quality, coarse threads, and heavy cloth for the colonial markets; and a printing and dyeing section which imported fine British cloth to produce finished textiles for the domestic and colonial markets.

Just as production was split so also were the interests of these two separate branches almost diametrically opposed. Printing and dyeing, which was tied more closely to British spinners and weavers than to those in Portugal, argued for low tariffs in order to import fine, undyed British cloth; spinning and weaving, on the other hand, desired high import tariffs at home and in the colonies in order to develop its manufacture of fine cloth and to protect its production of coarse cloth.[3] During the latter part of the nineteenth century, a number of fortuitous circumstances led to a burst of growth in spinning and weaving which increasingly challenged the political strength of the dyeing and printing branch. The period up to the 1920s saw frequent clashes between these two branches as they struggled to influence government policy.

THE STATE OF THE INDUSTRY: 1870–1910

Cotton textile manufacture in Portugal grew rapidly after 1870. An increase in domestic and colonial demand for textiles, as well as the recovery of American cotton production after the Civil War, attracted new entrepreneurs to the industry. Whereas in 1870 there were 50,060 spindles (including those still used domestically), there were 94,167 spindles ten years later according to the industrial census of 1881 (Table A13).[4] In spite of this impressive growth,

[3] For more information on the Portuguese textile industry during the 19th cent., see Pirio, 'Commerce, Industry and Empire'.

[4] H. P. Taveira, 'Memória', in Portugal, Ministério das Obras Publicas, Comércio e Indústria, *Inquérito Industrial*, 1881 (Part II, no. 1), 110–14 has a figure of 108,000 spindles for 1881.

however, the census showed that in addition to the split between printing and dyeing, and spinning and weaving, the industry was further dissected between cottage mills and factories, and manual versus mechanical looms.

According to the census of 1881, cottage mills rather than factories still dominated Portuguese textile manufacture. The 34 factories which were recorded only employed 5,747 people out of the total number of 39,597 people listed as employees of the textile industry.[5] The majority still worked in Portugal's substantial cottage industry supplying or competing against these factories or producing artisanal cloths. Moreover, the industry was falling behind technologically as evidenced by the high number of manual looms in the weaving sector. Mechanical or power looms using steam and then electricity were invented in Britain at the end of the eighteenth century and gradually replaced manual looms in British factories during the nineteenth century. In turn, the automatic loom, patented in the USA in 1894 and introduced in the UK in 1902 was a further improvement on the mechanical loom and subsequently replaced it during the early part of the twentieth century in most industrialized nations.[6] However, in Portugal, owing to lack of capital and the relatively cheap cost of labour, only 1,720 mechanical looms compared with 11,996 manual looms were in operation by 1881—again, most of these manual looms were probably in the cottage weaving mills.[7]

Despite the paucity of information on the cottage industry contained in the census, it gives us a glimpse of the shape and conditions of Portuguese textile manufacture in the late nineteenth century. The mills listed included old as well as new, large as well as small factories; and rich and poor ones. Tomar, one of the oldest factories in the country, was founded in 1789. By 1881 its capital was 400 contos and it employed 500 workers. Salgueiros, one of the youngest factories recorded, was founded a year before the census with an initial capital of 150 contos and had 363 workers. Lisbonense, founded in 1838, with capital of 500 contos already had 850 workers by 1881 whereas several cottage mills in Vila Real only had one or two weavers. The richest factory was probably the one at Braço de Prata, founded in 1876, which produced 460 contos

[5] Ibid. Part I, 56–8.

[6] Chris Aspin, *The Cotton Industry* (Aylesbury: Shire Publications, 1981; repr, 1987), 19–23.

[7] *Inquérito Industrial*, Part I, 58.

of textiles, a staggering amount when compared with 133 escudos earned by Joaquina Sancha, a weaver, working alone out of her home in the district of Vila Real.[8]

Regardless of size or age, working conditions were by all accounts appalling. Most factories worked from sunrise to sundown with only one and a half hours allowed for rest. Salaries varied from as much as 3$75 escudos per day for a master weaver to as little as 60 centavos per day for a minor under 16 years of age. Nearly all textile workers were illiterate and few factories provided schools or formal training for what was, except for artisanal weaving, a low skilled job. Of the 39,597 textile workers recorded in the 1881 census, 5,375 were women and 1,100 were listed as minors. Many of the women and minors were employed by the larger factories. The Companhia de Fábrico de Algodões at Xabregas which employed 51 minors out of a total of 191 workers, mentioned that one had begun work at 7 years of age. Salgueiros also had a 7 year old in its employment. Even those conducting the census exclaimed that: 'Certainly, in no other sector of national industry are workers more miserable.'[9]

In spite of the growth between 1870 and 1881, the Portuguese textile industry had not dislodged cheaper, better made British textiles from either the domestic or colonial markets. A succession of favourable tariffs, a more productive workforce, lower costs, and successful lobbying by the dyeing and printing section of Portuguese industry for the importation of finer cloths which Portuguese spinners and weavers did not make allowed Great Britain to retain a sizeable share of both markets, as it had done since the eighteenth century.

However, political pressure by textile manufacturers conspired with government priorities to promote the further development of the Portuguese industry in the late 1800s. Through their industrial organizations, the Associação Industrial Portuense (AIp—founded 1852) and Associação Industrial Portuguesa (AIP—founded 1875), spinners and weavers lobbied the Monarchy to secure the domestic and colonial markets for the industry at a time when the Portuguese economy was experiencing one of its periodic trade deficits. In response, the Monarchy passed a number of decrees in 1886 to help

[8] Ibid. Part II, nos. 1–2, for information on factories and working conditions.
[9] Ibid. Part II, no. 2, X, 137.

the industry capture more of both markets. But the real push came in 1892. Following the worldwide panic of 1890, the government devalued the escudo thereby increasing the cost of English textile imports and making domestic manufacture more attractive and more urgent. This was a bit of a double-edged sword as it also increased the cost of ginned cotton imports but nevertheless weavers and spinners began to move into production of finer threads and cloths at this time. Imports of English textiles dropped by 43 per cent between 1893 and 1898.[10]

The Portuguese government simultaneously passed protectionist legislation which virtually reserved the Angolan market for the textile industry. As a result, both branches of the industry prospered. This was effected by lowering customs tariffs on imports of metropolitan textiles and implementing a drawback scheme on cloth which was imported from Britain, printed and dyed in Portugal, and re-exported to Angola. A dramatic and immediate increase in exports occurred. Although cotton goods from the Congo free trade zone still managed to seep across the border, exports to Angola went from a paltry 131 tons in 1891 to a high of 3,894 tons in 1899 (Table A5). Similar measures could not be enacted in Mozambique due to Portugal's inability to control the contraband activities of Indian cotton textile traders operating on the border between the Portuguese colony and Southern and Northern Rhodesia.[11] Until well into the 1930s, Mozambique's textile imports were of British and Indian origin.

The Mozambican market notwithstanding, the capture of a share of the internal market and all of the Angolan one had startling effects on the growth of the industry, especially in spinning and weaving. Ginned cotton imports doubled from 8,232 tons to 16,360 tons between 1890 and 1910 (see Table A4). Many mills did so well out of the colonial market that they began producing textiles solely according to the tastes and demands of the colonies.[12] The number of spindles in the industry went from 300,000 in 1890 to 480,000 in 1913; and the number of mechanical (as opposed to manual) looms from 4,000 in 1890 to 15,000 in 1912.[13] Two of the largest

[10] Maria Filomena Mónica, *Artesãos e operários: Indústria, capitalismo, e classe operária em portugal (1870–1934)* (Lisbon: Instituto de Ciências Sociais, 1986), 160.

[11] Clarence-Smith, *The third Portuguese empire*, 91.

[12] Pirio, 'Commerce, Industry and Empire', 210.

[13] Mónica, *Artesãos*, 161.

factories, Salgueiros and Lisbonense, doubled their workforces in the 1890s and average profits of the large cotton mills reached as high as 19 per cent in 1898.[14]

Things went horribly wrong at the turn of the century. First, Portugal's former colony, Brazil, and Angola unintentionally provoked a reduction of textile exports to Angola. Brazilian over-production of coffee led to a glut on the world market causing a reduction in coffee prices. As coffee was one of Angola's two major exports, the price reduction greatly affected Angola's export income. Angola's other export, rubber, also dropped due to a decrease in world demand. As these two had accounted for 84 per cent of Angola's total exports in 1894, Angola's trade received a severe blow.

To make matters worse, a rise in the price of ginned cotton increased the cost of textiles further aggravating the situation in the Angolan market and causing disruptions in the Portuguese domestic market as well. To compensate for the loss of income from exports, the Angolan government had to reduce imports and Portuguese textiles were greatly affected. Textile exports from Portugal to Angola dropped from 3,894 tons in 1899 to 1,322 tons in 1901 to 981 tons in 1902 (Table A5). Total profits for the industry were only 15 contos in 1901 compared with 835 contos in 1898. Many firms collapsed.[15]

Though exports to Angola subsequently improved, the industry had come face to face with its dependence on the African market and its vulnerability with regard to ginned cotton prices. Dumping excess production on to the domestic market was a common but limited solution, since the coarse cloths produced for the colonies were unpopular in Portugal and fine cloths encountered cutthroat competition. After this experience, many of the larger spinning and weaving mills adopted the production of finer cloths, added printing and dyeing sections, diversified their output and in doing so, gradually began to overcome the division of needs and interests between the spinning and weaving section and the printing and dyeing branch. But integration did not happen overnight; it did not include everyone; and certainly, it was not immediately successful.

[14] Ibid. and 'Capitalistas e industriais (1870–1914), *Análise Social*, 23: 99 (1987-5), 837.

[15] Mónica, *Artesãos*, 162.

More typically, through its associations, the industry pressured the government to revise tariffs and protect markets, but there was little that the Monarchist government could do in the face of economic decline and political instability within Portugal.[16]

THE REPUBLICAN PERIOD

Unfortunately for textile manufacturers, the Republican government had even less interest or success in resolving their troubles either in the colonies or in the metropole. Exports to the colonies dropped from 3,864 tons in 1910 to 1,700 tons in 1911 and apart from a few exceptional years, they remained at or below that level right up to the 1930s (see Table A5). At the same time, the colonies were still importing on average 5,000 tons of textiles from foreign markets.

A drop in rubber prices after 1910 explains the initial drop in exports to Angola in 1911 but fails to explain why exports stagnated until the early 1930s. Certainly high prices and inflation disadvantaged Portuguese textiles in the colonial market. Moreover, shipping rates on Portuguese ships had increased during the war and this further contributed to the higher cost of Portuguese textiles. In addition, though metropolitan textiles enjoyed preferential tariffs in the colonial markets, duties for foreign textiles were not *ad valorem* but set prices per kilo. These were reduced by 25 per cent in 1912 and then not revised in line with inflation so foreign competitors were in effect not paying much more than Portuguese textile exporters by the mid-1920s.[17] Textile industrialists in this period bitterly complained about British competition in both Mozambique and Angola and through their associations, called on the government to revise tariffs.[18]

To cap the frustration of the industry in colonial markets, payments for goods imported into the colonies became paralysed in the beginning of 1924 owing to the colonies' poor financial state. Textile industrialists did not receive payment for goods exported to

[16] Ibid.

[17] Clarence-Smith, *The third Portuguese empire*, 119; and Companhia Rio Ave, *Relatório*, 1912, 8.

[18] *O Trabalho Nacional* (*OTN*), 9: 102 (June 1928), 1085–6.

the colonies and were then unable to pay the cost of raw materials, energy, or salaries.

With the colonial market stagnant, textile industrialists fell back on the only other available one, the domestic market. World War I briefly stimulated internal demand for the industry, but British textiles returned to compete with Portuguese ones after the war. In 1920 the share of foreign textiles in Portuguese imports was 12.83 per cent, even slightly higher than 1910 levels.[19] By 1926, the year the Republican government succumbed to a military takeover, Portugal was still importing 2,249 tons of textiles more cheaply from abroad. Thus not only were Portuguese textiles shut out of the colonial market but also they were competing with foreign imports in the domestic one. As in the past, too many textiles were chasing too few consumers. Severe competition arose sending the industry into crisis.

General economic and political conditions in Portugal exacerbated the industry's critical situation. Shortages and high costs of raw materials such as cotton or energy, the lack of credit due to the government's debilitating financial position, alternating policies of devaluation which cheapened exports but increased imports, and stabilization (which produced the opposite results) severely affected the rate and cost of production. Thus, even though spinners and weavers had moved into the production of finer cloths after the 1899 debacle, higher costs prevented them from competing successfully against the British. Secondly, as noted in Chapter 1, following the election of the Republican government, emigration increased dramatically, slackened during World War I, and then rose again after the war. Additionally, families rather than males began to emigrate from 1920 creating severe difficulties for the textile industry which employed both women and minors. According to official figures, moreover, most emigration was from the Northern region where the industry was concentrated, and resulted in severe labour shortages for mills. In the period from 1901 to 1930, total official emigration from the districts of Oporto and Braga, the locations of most textile mills in the country, was 118,501 and 64,506 people respectively (see map 1). The district of Oporto showed the second highest rate of emigration in Portugal from 1901 to 1930 and the Minho province, which included the

[19] Castro, *A economia*, 244.

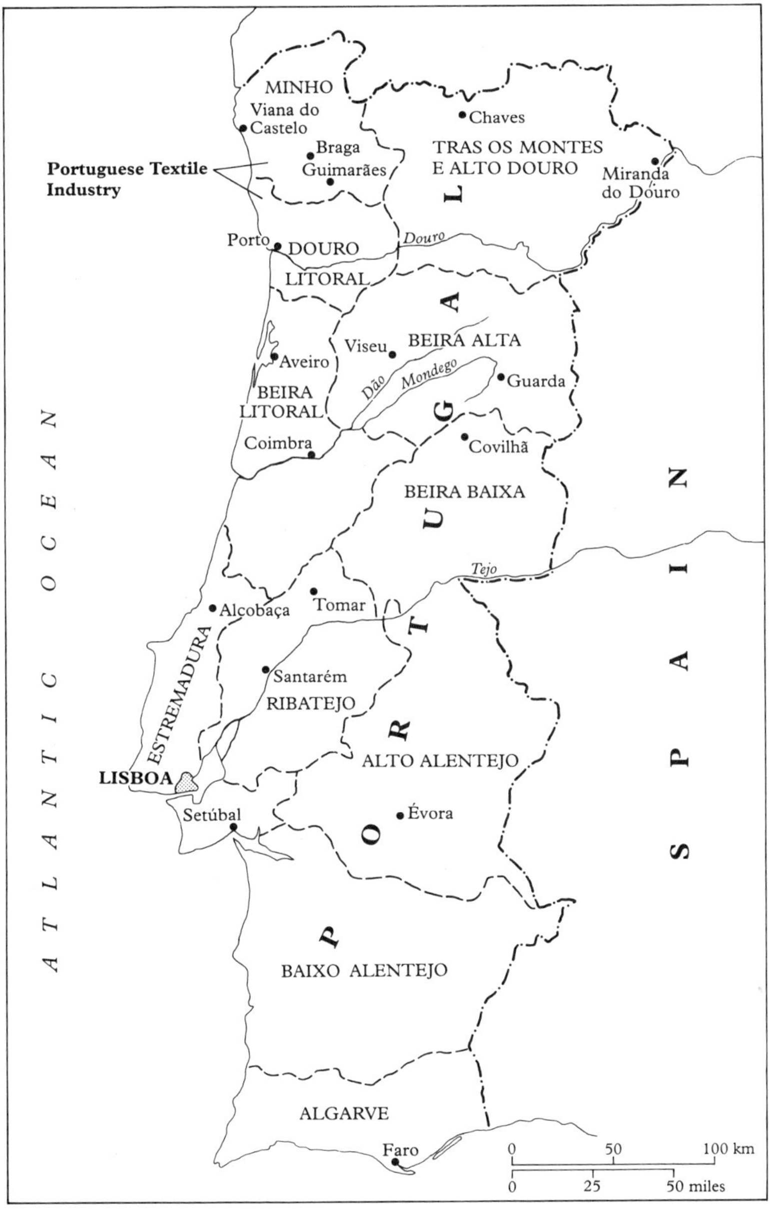

MAP 1. Portugal, Provinces, 1933. (Adapted from Douglas Wheeler, *Republican Portugal*.)

Note: Although officially part of the Douro Litoral province after 1933, popular usage included Oporto and its surrounding district in references to the 'Minho'. I adopt this practice in the text.

districts of Oporto, Braga, and Viana do Castelo, was the centre of emigration during this period.[20] Thirdly, related to the labour shortages as well as the general atmosphere of economic decline, were numerous labour strikes calling for better working conditions and wages. In the period after the war, 52 per cent of the strikes were related to wages and working hours.[21] Taken together with the specific problems of markets and competition, the general economic and political situation in Portugal worsened the position of the textile industry.

Company reports best illustrate the many difficulties of the industry and the inability of the government to solve them. In the initial years of the Republican government, both the Companhia Fiação e Tecidos do Porto and the Companhia Rio Ave referred to the high cost of raw materials, complained about the stagnation of the colonial market, and blamed the government for permitting foreign imports to flood the colonies as a result of low customs duties. By 1914 both cotton and the price of energy had increased and both companies mentioned the over-production and competition resulting internally as a result of the loss of colonial markets.[22]

At the start of World War I, the Companhia Fiação e Tecidos de Alcobaça reiterated these earlier complaints. In its statement for 1915, it said:

> The constant increase in prices of various primary materials, fuel, and manufacturing accessories (some of which attain outrageous levels, not to mention the impossibility of obtaining them at times) plus the increased labour which is needed to meet the expanded production [as a result of the war] are more than adequate reasons for us to be prudent and cautious in order to be prepared for any eventuality.[23]

By 1916 this company complained that machinery had begun to depreciate and could not be replaced due to the war. The following year, the company at Alcobaça again mentioned the excessively high price of cotton and the difficulty selling textiles because of

[20] Caroline Brettell, *Men who Migrate, Women who Wait* (Princeton, NJ: Princeton Univ. Press, 1986), 87.

[21] Marques, *History of Portugal*, ii. 138–9.

[22] Companhia Fiação e Tecidos do Porto, *Relatório e direcção; parecer do conselho fiscal*, 1912, 1914 (hereinafter Porto, *Relatório*).

[23] Companhia Fiação e Tecidos de Alcobaça, *Relatório da direcção e parecer do conselho fiscal*, 1915, no page (hereinafter Alcobaça, *Relatório*).

their high price. The Companhia Fiação e Tecidos do Porto also noted the high price of primary materials but was still able to find markets at home and abroad for its textiles.[24]

Partly as a result of military procurement, demand had actually increased during the war thus temporarily assuaging the fears of textile industrialists. The real troubles began after the war. Porto reported oscillating prices for fuel and primary materials in the first half of 1918 but had no trouble selling its output. Once the armistice had been signed, however, they immediately lost markets. Orders were cancelled and as a result production stagnated in the second half of 1918. The company reported: 'Seeing the signs of an armistice on the horizon, buyers dropped their purchases, and cancelled their orders, completely paralysing the placement of our merchandise.'[25] By November of 1918, the factory had to reduce working days.

The decrease in sales continued into the following year, and the difficulties of this company were compounded by a strike on 1 October 1919. Strikes were prevalent in the country after the war because of inflation, lay-offs, and poor working conditions; in this case the strike was called to enforce legislation that had already been passed by the Republican government, but ignored by this factory—the eight-hour day. Even though the Porto company was not working to capacity it still wanted its employees to work the traditional ten-hour shifts (two a day), rather than eight-hour shifts (three a day). Emigration had been increasing after the war and the company feared labour shortages if it switched to a three-shift system.[26] In its company report for 1919, the company bemoaned the labour shortage and berated the government for passing an eight-hour law despite criticism by industrialists:

> It was a grave economic error for the country that eight hours of work were decreed for all industry; the shortage of workers in all industrial branches was ignored as were the just representations made by commercial and industrial associations of the country with respect to that odious decree.[27]

[24] Porto, *Relatório* for 1915 and 1917, shows a rise in the cost of total cotton purchases from 72,924$61 in 1915 to 126,885$32 in 1917 as a result of the rise in international cotton prices and inflation. See Table A8 and Fig. A2 for real world prices after 1914.

[25] Porto, *Relatório*, 1918, no page.

[26] Ibid. 1919, 7.

[27] Ibid.

The problems did not abate in the early 1920s. Costs of primary materials continued to oscillate and coupled with the inflation, strikes, and financial insecurity at home, textile industrialists found themselves in an intolerable situation. Fierce competition continued in the internal market and exports to the colonies remained stagnant in this period. By 1924, with the stabilization policy and the paralysis of transfers compounding the difficulties of the industry, some factories were working only two or three days a week.[28] The situation in the colonies was particularly singled out by the Porto company which declared in its 1925 statement: 'The lack of a solution to the colonial problem is the prime factor in the critical situation of the cotton industry.'[29]

With the fortunes of Republican governments oscillating as much as the markets in these years, consistent government action was impossible. The colonies were virtually ignored as successive governments struggled with deteriorating economic and political conditions at home. Emigration, worker unrest, unwelcome legislation, erratic prices, unsettled markets, and a swelling opposition to government policy suggest that textile industrialists, like many others in Portugal, no doubt welcomed the 1926 coup.[30]

1926–1936: CONTINUITY AND CHANGE

Immediately after the coup, the Associação Industrial Portuguesa and the Associação Industrial Portuense, in a letter to the Minister of Finance, expressed their desire 'to contribute to the success of the program of national reconstruction being attempted by the government and most especially, the Minister'.[31] Repudiating the failures of the old goverment, they then suggested measures which could help industry recover lost markets and improve its position in the Portuguese economy. These included the reform of customs duties, the reduction of taxes, the abolition of *ad valorem* taxes, and provisions for the availability of credit to industry. They disingenuously added that 'the interests of industry are joined with

[28] Medeiros, *A sociedade*, 377, see n. 10.
[29] Porto, *Relatório*, 1925, 5.
[30] See Telo, *Decadencia*, 367, on the shift to the opposition by textile industrialists.
[31] *OTN* 7: 79 (July 1926), 682.

those of the Nation and the State, it is not aiming for special favours'.[32]

With the blessing of industrialists, then, the government immediately passed a decree encouraging the development of cotton in the colonies but exports were another matter. In the late 1920s only about one-fifth of textile imports into the colonies were Portuguese; the rest were British or Japanese, and the industry cited this disparity as the cause of the severe crisis from which it was suffering. As usual, the industry was keen to improve its own exports and engaged in a big lobbying effort to get the government to offer protection to Portuguese textiles.[33] In 1928, 150 representatives of the industry met with the Ministers of Finance and the Colonies to argue for a greater share of colonial markets. They stated that many factories had reduced working hours and others had shut due to the lack of sufficient demand for Portuguese textiles. The colonial market was seen as the major cause: 'For years and years, cotton industrialists have vainly called on the government to find markets for national products within the Portuguese dominions, especially in Angola.'[34]

Angola was cited as having imported national, undyed and white cloth worth 266 contos and identical foreign cloth worth 31,180 contos in 1925. Moreover it had imported foreign dyed cloth worth 83,344 contos, while its imports of national dyed cloth had only amounted to 4,665 contos.[35] Textile industrialists argued that if the government were to reinstate the 1892 tariffs which required the payment of colonial customs duties in gold by foreigners, then the textile industry could capture this market again.

But if textile manufacturers thought the new government was going to give priority to securing markets in the colonies, they were soon disappointed. Initially, the new government was both unwilling and unable to satisfy the industry's need for external markets. One of the reasons for this was simply that it did not have sufficient control in either colony to enforce any of its policies (as the next chapter illustrates). The government therefore had to tread

[32] Ibid. 683.

[33] See *OTN* 1926–36: whereas the textile industry frequently lobbied the government to find markets, it hardly ever referred to problems with the cotton supply except in the years when prices were high, as we have mentioned.

[34] *OTN* 9: 102 (June 1928), 1085.

[35] Ibid. 1086.

carefully. The Angolan administration was strongly influenced by settlers so the government had to balance textile interests against settler support. Mozambique was still dominated by concessionary companies and the Portuguese government barely controlled the frontiers of that colony, a situation which prevailed until the early 1930s.

The Angolan market was crucial for textile industrialists. Approximately forty firms were said to have been dependent on exports to the colonies—no doubt as a result of their earlier connection.[36] They included firms such as the Companhia da Fábrica de Fiação de Tomar, Companhia de Fiação e Tecidos de Fafe, Companhia de Fiação e Tecidos de Guimarães, Companhia Fabril de Salgueiros, Companhia Fiação e Tecidos de Alcobaça, and Companhia Fiação e Tecidos do Porto, many of which we have already mentioned and will mention in this study. Most of these firms were now amongst the most powerful in the industry; they were the most active in the industrial associations and tended to dominate meetings. They were often behind the letters and telegrams sent to government ministers and campaigned fiercely for favourable customs duties for Portuguese textiles, high tariffs for foreign imports, and the prevention of colonial textile industries. For example, in 1928, the Companhia de Amboim requested permission to set up a textile factory in the colony to produce textiles for the whole of Angola. This company, whose main activity was coffee production, had recently been rescued from financial collapse by a Belgian group.[37] The takeover and request followed an agreement between the Belgian Congo and Angola to co-operate on matters of trade, tariffs, communications, and other areas. The request conflicted with a 1914 decree prohibiting the construction of any industry in the colonies which competed with a Portuguese domestic industry.[38] It was probably an attempt to test whether the new Portuguese government would uphold previous legislation or be willing to use the recent agreement to get around the 1914 decree.

The domestic textile industry reacted strongly to the request. In a letter to the government, cotton industrialists complained that

[36] *OTN* 16: 180 (June 1935), 2454.
[37] Clarence-Smith, *The third Portuguese empire*, 171.
[38] *OTN* 9: 105 (July 1928), 1101–3.

two months earlier they had asked the government for help with the Angolan market. They had never dreamed that help would come in the form of a request to set up cotton manufacture in Angola which,

> aims to rob metropolitan industry of the possibility of recapturing the Angolan market, as requested of the government, and of enlarging it [the market] as much as possible in order to secure the prosperity of the metropole and assure the well-being of a great part of the Portuguese family.[39]

A year elapsed and the government had not responded to the complaints of textile manufacturers. Through the AIp, textile manufacturers took their complaints to the High Commissioner of Angola instead of the Minister for the Colonies, presumably because they felt the Minister did not have sufficient control over the colony. They asked that Portuguese industry be given a 50 per cent reduction on customs duties, that specific taxes rather than *ad valorem* duties be levied on Portuguese goods, and that greater vigilance be exercised on the border with Northern Rhodesia, through which, they claimed, much contraband passed.[40] By 1930 customs duties had still not been changed sufficiently to make a difference nor had any action been taken on the Amboim request. Angolan authorities claimed that metropolitan industry did not understand the situation in Angola. The industry responded that foreigners in Angola were pressuring Angolan authorities to keep duties down. Despite repeated telegrams sent by the AIp to the central government little action was taken until 1931.

The protests against the establishment of colonial industries, the pleas for the imposition of customs duties favouring Portuguese textiles, and the flip-flopping between the High Commissioner and the metropolitan government are a specific illustration of the general climate at the time. Colonial control was still beyond the power of Lisbon; Lisbon was not yet able to impose its economic policy either in the colonies or at home in the late 1920s and early 1930s. Instead, as illustrated in Chapter 1, the metropolitan government appeared to be manœuvring amongst the various pressure groups, colonial and domestic, trying to balance interests.

[39] *OTN* 9: 104 (Aug. 1928), 118.
[40] *OTN* 10: 115 (July 1929), 1296–7.

A similar pattern of ineffective lobbying also initially took place in the metropole. Although the rate of growth in textiles was lower than that of other industrial sectors for the period from 1917 to 1930 (Table A3), national industry claimed that severe competition amongst companies was taking place in the internal market due to the loss of the colonial markets and due to continued imports of foreign goods. As a result the industry declared that it was in 'crisis' and suggested solutions to save it. The first of these was to improve conditions in the industry; the other, to control its growth and activity.

In 1926 metropolitan Portugal was still importing 1,928 tons of foreign textiles and 321 tons of thread from abroad.[41] While the textile industry was keen to replace these goods with its own, it was also very worried about the internal dumping that occurred whenever the markets contracted. What several firms within the industry wanted was a way of controlling the industry to prevent overproduction and therefore cutthroat competition. Amongst the several solutions proposed was the compliance with the eight-hour working day legislation passed during the Republican government and prohibitions on purchases of new machinery, the construction of new factories, or the expansion of existing ones.

The request for compliance with the eight-hour day legislation was ironic considering the industry's earlier protests over working hours but the labour situation and the political environment had changed significantly since the Republican period. First of all, with all countries experiencing high unemployment following the Depression, emigration from Portugal slowed. The official average yearly rate of emigration for the 1930s was 11,918 emigrants per year as compared with 36,634 emigrants per year for the 1920s.[42] Faced with the prospect of high unemployment in Portugal and its attendant social consequences, both the military government and the *Estado Novo* were probably reluctant to change the eight-hour law, especially if retaining it meant added employment in those companies which operated shifts. Secondly, acknowledgement of the eight-hour legislation was a concession to labour, the carrot to the stick of labour repression. Faced then with an inalterable situation, several of the larger textile industrialists decided to make

[41] *Indústria Têxtil*, 1: 5 (Apr.–May 1937), 11.
[42] Serrão, *A emigração*, 31.

sure it was enforced in the entire industry. In 1930 through the AIp, twenty-six of the largest textile factories sent a letter to the ministries of the Interior, Finance, and Commerce and to the Director-General of Industry stating that small (probably cottage) mills in rural areas were not complying with the eight-hour legislation and were working sixteen-hour days. These factories were also paying far lower salaries and taxes with the result that they were not only over-producing but also selling at unreasonably low prices, resulting in fierce competition in the domestic market.[43] Three months later on 20 October 1930, the AIp followed this letter with another request to the same ministries. It stated that some factories were working only three days a week and that in addition to finding markets, the government should institute regular working hours and force companies to comply with them.[44] A month later, claiming that the industry was already oversaturated with looms and over-producing, the cotton section of AIp again asked the Minister of Commerce to turn down a request for the purchase of ninety looms made by three weaving establishments.[45]

In the early 1930s the government started to take legislative action to secure markets and to control competition. While the lobbying efforts of the industry probably influenced the content of the decrees, it is likely that the Depression as well as Salazar's own dominance of economic policy by this time dictated the nature of decree laws relating to the textile industry as they had done with other Portuguese industries. Although Portugal was probably less dependent on the world market than other countries, the wave of protectionism which swept the industrialized countries gave Salazar the opportunity to protect metropolitan trade with the colonies and thereby to mitigate the effects of the Depression for Portuguese industries, notably wine and textiles. Establishing control over trade with the colonies allowed the government to improve the metropole's balance of trade and provided outlets for Portuguese industry which thereby contributed to political stability. Controlling trade then was a way of controlling the political arena. Fortuitously, the world Depression gave the government the excuse to intervene further in the economy and to take an autarkic path in

[43] *OTN* 11: 128 (Aug. 1930), 1501–2.
[44] *OTN* 11: 131 (Nov. 1930), 1555–6.
[45] *OTN* 11: 132 (Dec. 1930), 1572.

the colonies without the retaliation that normally occurred when countries engaged in protectionism.

In 1931 the Minister for the Colonies turned down the request for the textile factory in Angola by the Companhia de Amboim. This was then followed by Decree-Law 19,773 which reorganized the system of currency transfers between Angola and Portugal allowing greater facility and speed in currency exchanges between the two countries. Thirdly, customs duties were revised to favour the textile industry. This was strengthened in 1932 by the addition of a 15 per cent tax on foreign textiles imported into the colonies. The same decree prohibited the creation of colonial textile factories.[46]

Domestically, in February of 1931, the Ministry of Finance published instructions for compliance with the eight-hour working day legislation.[47] This was followed in 1933 by comprehensive labour legislation governing working conditions, wages, and employer–employee relations, though in practice the government only half-heartedly enforced these regulations. Furthermore, the Minister of Commerce announced a census for the industry to ascertain its size and its requirements. The last complete census had been in 1881 and industrialists had been requesting a new one for years.[48] In addition, customs duties on imports of foreign textiles into the metropole were increased. Finally, and most importantly, the Ministry of Agriculture, Commerce and Industry published comprehensive legislation, Condicionamento Industrial (Industrial Regulation), to regulate the growth and expansion of the industry. For textiles, all firms except those which used less than five workers or under 5 horsepower, or were producing for household or artisanal purposes—the cottage mills—were subject to the legislation. Having complained repeatedly about over-production and stiff competition, those industrialists represented by the AIp, which included the largest and most influential textile firms, welcomed such legislation though they were no doubt disappointed at the exclusion of cottage mills, which they partially blamed for the cutthroat domestic competition.

The ever-increasing encroachment of government intervention

[46] Portugal, *Diário do Governo*, Série I, D-L 21,226 (22 Apr. 1932).

[47] *OTN* 12: 134 (Feb. 1931), 1604.

[48] See *OTN* 9: 108 (Dec. 1928), 1181–3, for complaints about the lack of a census and government response, *OTN* 12: 135 (Mar. 1931), 1626.

into the textile industry was greatly advanced by the passage of the Industrial Regulation law. As we mentioned in the previous chapter, the regulations applied to a majority of Portuguese industries at the time and intended to protect industries by limiting the construction of new ones, and the expansion of old ones. At the time, businessmen failed to perceive that the legislation would result in wholesale government meddling in industrial affairs. In the midst of the Depression, industry was weak and welcomed government interest. Most thought it temporary.[49]

For the textile industry, it probably came as a welcome sign of government intentions to help, although subsequently the decree was both less effective and more constricting than even the industry would have wanted. It was a compromise. In return for slow growth and government control, the industry would be protected. In return for the colonial market, the industry agreed to government regulation. This was the package constructed by the *Estado Novo*: slow growth, control of competition, and protected markets under government supervision.

EFFECTS OF LEGISLATION

The reaction to and results of the industrial and commercial legislation were mixed. Portugal's exports to the colonies doubled from 1,298 tons in 1929 to 2,950 tons in 1935 (Table A5). In Mozambique, a market which had long eluded the Portuguese textile industry, the industry's share of textile imports went from 4 per cent in 1929 to 55 per cent in 1935. In Angola, 85 per cent of total textile imports came from the metropole in 1933.

Portuguese textiles had success also in the internal market. In 1926 imports of foreign cotton goods equalled 2,462 tons; by 1935 foreign imports had dropped to 901 tons.[50] Protectionism completed the integration between the spinning and weaving, and printing and dyeing branches of the industry but without reducing the fierce competition among mills. The greater the share of the colonial and domestic markets the industry captured, the more

[49] Brito, 'Corporativismo', 54–6.

[50] Freppel Cotta, *Economic Planning in Corporative Portugal* (London: P. S. King & Son, Ltd., 1937), 136.

competition seemed to increase.[51] During the 1930s the Guimarães company repeatedly referred to domestic and colonial competition because of constant changes in raw material prices, fluctuating exchanges, and the opening of new factories, as did the Fafe company which actually called for a study of the needs of the market in relation to the state of the industry. In 1934 the mill at Alcobaça complained bitterly that:

> During the six months which just ended, everything passed normally, nevertheless, certain businesses engaged in unfair competition which resulted in a loss of profits on our sales.[52]

Two reasons explain why competition did not decrease in this period. First, Industrial Regulation exempted cottage mills, which were still numerous. Through the 'putting out' system whereby larger factories subcontracted work out to cottage mills, these home industries still commanded a share of textile production. Secondly, of those mills covered by the regulations, many expanded enormously in spite of the legislation. The factory at Guimarães noted that it received ministerial authorization to buy twenty-three looms in 1933, in the midst of the Depression, and in subsequent years the company report noted with satisfaction that the factory was replacing equipment. This seemed to have been what other companies had in mind too because in 1933 the cotton section through the AIp requested that no new factories be built and that requests for new machinery be turned down, since it was evident that the industry had already saturated available markets. The next year the industry protested strongly to the Direcção Geral da Indústria (DGI) not to grant authorizations which had been requested for 40,000 new spindles.[53]

The following year the factory at Guimarães warned that the construction of more and more factories in the country would create serious difficulties for the industry. It was over-producing already and the price of textiles was dropping:

> It suffices to note that consequently, many industrialists recently have been frightened by the existence of a great number of manufactured articles. Principally, these articles do not require great technical skill to make.

[51] *Indústria Têxtil*, 1: 5 (Apr.–May 1937), 1.
[52] Alcobaça, *Relatório*, 1934, no page.
[53] *OTN* 16: 186 (June 1935), 2459.

Everyone is driven to find placement for them and when they do not find it easily, they dump the articles on the markets at unbelievable prices. They have been acting truly insane.[54]

One state official later admitted that bureaucracy at the time was inexperienced, statistics and studies of industry generally were unavailable and that as a result, the policy of Industrial Regulation was not well executed.[55] Since no complete census had been taken of the industry since 1881, statistics were compiled by different organizations and were often conflicting at least until the creation of the Comissão Reguladora do Comércio de Algodão em Rama (CRCAR) in the late 1930s. Because of the unreliability of statistics, it is difficult to say what kind of growth the industry experienced or when it occurred. But those available statistics as well as cotton imports may give an approximation. Ginned cotton imports stood at 16,360 metric tons in 1910. In 1913, according to Indústria Portuguesa, the journal of the Associação Industrial Portuguesa, the textile industry had 480,000 spindles (see Table A13). The next available reliable figure cited the textile industry as having 600,426 spindles in 1935. A growth of 120,000 spindles between 1913 and 1935 seems remarkable in light of the difficulties encountered by the industry after the war and the Depression. It is important then to pinpoint the period of growth.

In addition to stimulating demand, World War I reduced supplies of British cloth forcing the dyeing and printing branch to rely increasingly on domestic cloth production, so it is likely that the number of factories and spindles rose to meet the added demand. However, with the return of British imports, the loss of markets, and high inflation after the war, it is probable that many of the factories temporarily shut down or reduced work, hence the small increase to 503,000 spindles attributed to the industry in 1928. After 1931 the existence of protected markets probably led to the reopening or growth of many factories or to greater use of spindle capacity. Moreover, quoted imports of raw cotton equal to 22,404 metric tons in 1932 would seem to suggest that protectionism completed the integration of the spinning and weaving, printing

[54] Companhia de Fiação e Tecidos de Guimarães, *Relatório da direcção, balanço e parecer do conselho fiscal*, 1935, 8 (hereinafter Guimarães, *Relatório*).

[55] J. D. de Rego Lima, *O Condicionamento Industrial: Instrumento de reorganização* (Lisbon: Tip. da Empresa Nacional de Publicidade, 1966), 29.

and dyeing branches. It appears that as soon as markets were captured, however, more mills reopened, were founded, or worked to capacity in order to take advantage of the markets. As the industry was already competing in saturated markets, the Industrial Regulation legislation was supposed to avoid further expansion and growth, but as company criticisms suggest, it was unsuccessful. By 1935 there were 600,426 spindles in the industry and raw cotton imports stood at 22,601 metric tons. This represented an increase of between 100,000 and 120,000 spindles since 1913, *and* 1928, since many had lain dormant during the crisis. In spite of government intentions to reduce competition in the industry, Industrial Regulation had done little to control it even during a period of crisis. In effect, among cottage mills, the government did not know what it was regulating; among factories, it was not enforcing the regulations very well. Little wonder then that companies continued to complain about metropolitan and colonial competition even after capturing a large share of both and repeatedly, to demand a census.[56] Now they were competing against each other.

We turn now to the third partner in the triangle that was being constructed in the 1930s. This partner was not a cipher waiting to be fitted into the designs of the metropolitan government and the textile industry, but an entity with a past and with interests and contradictions of its own. An examination of the situation in the colonies will explain why the cotton regime was not an initial success nor markets immediately guaranteed.

[56] See *OTN* 12: 143 (Nov. 1931), 1744—industrialists complain about the creation of new factories. Further complaints about the lack of a census can be found in *OTN* 14: 159 (Mar. 1933) and 17: 118 (June 1936).

3

The Establishment of the Cotton-Growing Regime

The period from 1926 to 1936 was a period of reorganization in the colonies as it was in the metropole. First the military government and then the *Estado Novo* initiated colonial programmes of administrative overhaul and economic reform. These policies were not initially successful but the intent was clear: the colonies were to benefit the metropole and the state would intervene to force compliance. The creation and subsequent development of the cotton regime embodies some of the central tenets of these general policies and illustrates some challenges faced by the new regime. The passage of the cotton decree in 1926, a fixed and guaranteed price, bonuses for production, and the establishment of concessions testify to a regime committed to autarky but not yet possessing the strength to secure it.

THE 1926 DECREE AND OBSTACLES TO ITS IMPLEMENTATION

Decree-Law 11,994 of July 1926 has been cited as the hallmark of the Salazar regime and indeed of Portuguese colonialism. The decree established a compulsory cotton-growing scheme using African labour in the Portuguese colonies and it remained virtually unchanged until the 1960s. However, the schemes of the military dictatorship and the *Estado Novo* were not new: legislation as well as attempts to grow cotton had gone before under the Monarchy and the Republican government.

Since I have discussed elsewhere the endeavours to grow colonial cotton by the Monarchy and Republican governments, a brief summary will suffice here.[1] In spite of the fact that cotton legislation

[1] M. Anne Pitcher, 'Sowing the Seeds of Failure: Early Portuguese Cotton Cultivation in Angola and Mozambique, 1820–1926', *Journal of Southern African Studies*, 17: 1 (Mar. 1991), 43–70.

dates back at least to the time of the Marques de Pombal, neither the Monarchy nor the Republican governments were able or interested in promoting sustained cotton production in either Angola or Mozambique. Natural causes such as pests, poor soil, and unsuitable climate contributed to their poor record but erratic international prices, conflicting and contradictory metropolitan objectives in the colonies, inappropriate methods of production, competing crop and labour demands, and African resistance all had their part to play in the failure. In the history of previous interest in the crop, the highest recorded output in Mozambique and Angola was 2,744 tons and that, in the final year of the Republican regime.

Yet from the ashes of earlier cotton-growing attempts, the military dictatorship and the *Estado Novo* which followed drew key tenets of their cotton policy. Experience in Mozambique and legislation passed by Governor Couceiro in Angola under the Monarchy were direct precursors of the coercion, use of peasant and female labour, and price-fixing adopted by the *Estado Novo*. Where earlier attempts to employ the plantation system to grow cotton had failed, the new regime adopted forced peasant production, based on the Belgian model but derived equally from their own experience. Where international prices had previously determined increases and decreases in output, the new government guaranteed the price; and where earlier governments had left companies and settlers in control of cotton production, the new government adopted direct state intervention.

Decree 11,994 was passed on 30 July 1926 only two months after the coup. Swift legislative action with regard to colonial cotton production suggests early pressure on the new government by textile industrialists but the government itself had more than sufficient reason to foster cotton cultivation. The year before the decree, the government had a negative balance of trade of 1,621,973 contos.[2] Ginned cotton only accounted for 7.4 per cent of the total value of imports in 1925 but this was an important margin in terms of the overall deficit and, as opposed to machinery or coal, cotton was an import which could be substituted by colonial production and controlled by the government. The state made this quite clear in the decree, by referring pointedly to the negative effect of cotton

[2] Rosas, *Nos anos trinta*, 67.

imports on the country's commercial balance. Cotton was quoted as costing Portugal approximately 150,000 contos a year 'with grave effects on the national economy'.[3] Trade statistics quoted an even higher figure for cotton imports in 1925: 185,225 contos.[4] The government therefore had clearly earmarked it as an import which could be replaced by colonial production and not surprisingly, cotton became one of the regime's top priorities.

The decree was a product of its time and of previous experience. Its 54 Articles addressed everything from the type of cotton seeds to be used to the method of cultivation. It not only drew elements from previous decrees but copied many of its colonial neighbours in approach as they, in turn, copied Portugal. Imperial governments in other European countries had tried and were trying to institute similar cotton-growing systems in Africa. In fact, Decree-Law 11,994 of 1926 referred to the 'brilliant results obtained by the English, French, and Belgians' as justification for revived Portuguese attempts to grow cotton in Mozambique and Angola.[5] Prior to World War I, the Belgians had studied British, French, German, and Portuguese Agricultural Reports and visited the Sudan, Uganda, and Mozambique before adopting a system based on forced peasant production in 1917. After poor results with plantation production of cotton, Portugal, like Belgium, advocated legally enforced, compulsory cotton cultivation by African peasants and sale to cotton concessionary companies who had monopoly buying, ginning, and exporting rights within specially demarcated zones. Like Britain, all trading and ginning activities were reserved to Europeans to prevent Indians and Africans from organizing it themselves. Finally, like all other imperial powers, the decree called for active government supervision of the entire production process from seeding to shipping. What was unique about the Portuguese approach was its longevity. State supervision and the legal compulsion to work lasted into the 1960s, long after Britain and Belgium had dropped their cotton schemes.

The 1926 decree intended to capitalize on favourable circumstances created by an increase in the world price of cotton which had begun during the Republican period. In the past, increases in

[3] *Diário do Governo*, Serié I, D-L 11,994 (30 June 1926).
[4] Rosas, *Nos anos trinta*, 70.
[5] *Diário do Governo*, Serié I, D-L 11,994 (30 June 1926).

the world price of cotton produced jumps in cotton output from Portuguese Africa as settlers and companies tried to profit from high prices. The Republican period was no exception. The international price of ginned cotton had begun to rise from 1909, continuing to increase during World War I to reach the astronomical price of 19$80 a kilo in 1924. The high price was partly due to galloping inflation which occurred with the onset of World War I, but was also the result of a cotton shortage brought on by the war and a boll-weevil attack on the American cotton crop in 1923–4, which forced up the world price (see Tables A7–A8 and Fig. A2 on nominal and real prices).

The increase in cotton prices, coupled with the Republicans' gold duties on imports, elicited bitter complaints about high prices from textile companies, and prompted the Lisbon Geographical Society, a pro-colonial lobbying organization, to revive pleas for the creation of a cotton centre based on Britain's Empire Cotton Corporation.[6] Although no centre was begun, the rise in prices did prompt the founding of a few colonial cotton companies as had happened in the 1860s and early 1900s when shortages of supply had produced jumps in price. According to the Direcção dos Serviços de Agricultura (DSA): 'Large amounts of capital were invested in this area of agriculture and extensive areas were devoted to its cultivation.'[7] The *Empire Cotton Growing Review* of Great Britain remarked: 'Several cotton plantations have started operations in the Lourenço-Marques district, and native grown cotton is an important industry in the Territory of the Mozambique Company.'[8]

The Anglo-East African Cotton Company (registered in London, 1925) was reported to have acquired 75,000 acres in Inhambane and Premier Cotton Estates Ltd., also English, another 70,000 acres along the Lourenço-Marques and Ressano Garcia railway for the purposes of cotton growing, but this company collapsed within

[6] Porto, *Relatório*, 1923 and 1924; Sociedade de Geografia de Lisboa, *Téses e actas das sessões*, Segundo Congresso Colonial Nacional, 6–10 May 1924 (Lisbon: Tip. e Papel. America, 1924), 14.

[7] Moçambique, Direcção dos Serviços de Agricultura, *Algodão*, Colónia de Moçambique, Primeira Exposição Colonial Portuguesa, Oporto (Lourenço Marques: Imprensa Nacional, 1934), 10 (hereinafter Mozambique, DSA, *Algodão*).

[8] From *Tropical Agriculture*, 6: 2 (1929), 266 quoted in *Empire Cotton Growing Review*, 6: 1 (Jan. 1929), 52.

several years of its founding.[9] Alongside its other commitments, the Mozambique Company, a concessionary company dating from the nineteenth century also increased cotton production in the areas of Manica and Sofala, a region where it had experimented with cotton growing since 1904.[10] GRANDUCOL, a company financed with capital from Luxemburg, had also acquired concessions in the Zambezia region by 1920.[11]

Especially from 1924, production in the state-administered territories of Mozambique began to increase quite rapidly. Exports of ginned cotton to Portugal and abroad went from 207 tons in 1920 to 226 tons in 1922 to 386 tons in 1924, reaching a high of 1,154 tons in 1926 the very year that detailed cotton plans were passed by the military government. Compared to what had gone before this was quite impressive. Moreover, in 1924 the first year for which figures from the whole of Mozambique are available, 742 tons of ginned cotton were exported from the territory of the Mozambique Company. Even the hopelessly managed Niassa Company in northern Mozambique managed to export 129 tons in 1924 reaching a high of 255 tons in 1929, the year their charter expired.[12]

In Angola, moreover, several companies and settlers also attempted cotton growing as a result of the price rise in the mid-1920s. In Angola the Fomento Geral de Angola was formed in 1920 with 522 contos in capital, rising to 5,400 contos in 1921, most of it supplied by its majority shareholder, the Belgian company, Société Générale de Belgique. This company encouraged cotton growing but pursued other agricultural interests as well. Also in operation by 1926 was the Companhia Geral dos Algodões de Angola which cultivated cotton in the area of Malange. The majority shareholder was again Société Générale with links to three Portuguese and two Belgian banks: the Banco Burnay, the Banco Nacional Ultramarino, and the Banco Comercial de Lisboa in Portugal; and the Banque Josse Allard and Banque Belge d'Afrique in Belgium; two

[9] C. Carlyle-Gall (ed.), *East African Manual* (London: Mining and Industrial Publications of Africa, Ltd., 1929), 866; and Carlos Fortuna, 'A mudança de côr do algodão moçambicano: Estado, capital e trabalho no período de entre-guerras', *Revista internacional de estudos africanos*, nos. 10–11 (Jan.–Dec. 1989), 109.

[10] Pitcher, 'Sowing the Seeds', 59.

[11] Castro, *A economia*, 118.

[12] Mozambique, DSA, *Algodão*, 6.

companies: the Companhia do Boror and the Companhia Industrial de Portugal e Colónias; and individual shareholders. Cotonang as it was called had an initial capital of 6,000 contos and subsequently became the largest and most profitable cotton company in Angola.[13] Ginned cotton exports from Angola showed similar jumps rising from 177 tons in 1920 to 674 tons in 1922 to 750 tons in 1926.[14]

Though production and exports were actually rising up to 1926, the Republican government still had no control over the destination of exports. Particularly in Mozambique, most cotton exports went to Britain.[15] Of the cotton exports which did go to Portugal, the total amount was meagre in comparison with the overall needs of the textile industry. Manufacturers required 15,000 tons annually. A mere 1,000 tons from the colonies really made very little difference and underscored the point that the Republican regime had done and could do little to improve production or exports.

Any hope that the new regime could improve this production or divert it to Portugal through the 1926 decree was quickly dashed. That year, international prices peaked and began to fall. The following year, total output in both Mozambique and Angola dropped in response to the drop in international prices. From 14$73 in 1925, the price dropped to 9$60 in 1926, to 9$05 in 1930, and finally to a low of 5$68 in 1931 (Table A7). Total output dropped to 1,953 tons in 1927 and stagnated at around 2,000 tons up to 1931. Exports to Portugal were under 1,000 tons for the same period.[16] For nearly six years the decree remained a dead letter.

From colonial company reluctance to the ambivalence of the metropolitan textile industry, a host of reasons illustrates why the policy was not immediately effective. In the colonies, particularly Mozambique, the drop in the international price of cotton, a negative report on Mozambique's cotton potential, a shortage

[13] See Portugal, Ministério das Colónias, *Anuário Colonial*, 1927–9 (Lisbon: Imprensa Nacional, 1930), 162.

[14] Great Britain, Board of Overseas Trade, *Report on the Economic Conditions in Angola* (London: HMSO, 1923–9). Figures have been converted to metric tons.

[15] Great Britain, Board of Overseas Trade, *Report on Economic Conditions in Portuguese East Africa* (London: HMSO, 1927–33).

[16] For total output, see Great Britain, Board of Overseas Trade, *Angola*; and Mozambique, DSA, *Algodão*. Owing to widely varying statistics on Portuguese imports of colonial cotton, these figures were approximated from Great Britain, Board of Overseas Trade, *Angola*; Mozambique, DSA, *Algodão*;; and Portugal, INE, *Estatísticas do Comércio Externo*, various years.

of labour, and lack of government control explain the negative response by companies and African producers to the 1926 decree. Because of the drop in the international price from 1926, many companies in Mozambique and Angola, fearing the risks involved, did not seize the opportunity to switch to cotton and continued other agricultural pursuits. In addition, an American cotton expert concluded in 1929 that Mozambique was 'marginal or sub-marginal' land for cotton growing. He found the uncertain rainfall and high prospects of cotton disease in Mozambique sufficient reasons to suggest that cotton growing was highly unlikely except for specified areas within each district.[17] The report further dissuaded some established companies from switching to cotton and probably frightened new companies from locating in the Portuguese colonies.

Furthermore, although radical alterations in colonial policy and administrative reorganization swiftly followed the coup in 1926, the government still exercised little effective control in the colonies, even in the section supposedly administered directly by the state.[18] Moreover, in the Manica and Sofala region of Mozambique, because the Mozambique Company was a type of concession which had state as well as economic functions, the Portuguese government had little leverage in trying to force this company to continue or increase cotton production if it did not want to, or to sell to Portugal. In the years immediately after the passage of the legislation, the Mozambique Company as well as companies in the state-administered territory continued to sell the majority of their cotton to Great Britain.[19]

As for the northern part of Mozambique, it too was under charter to a British concession, the Niassa Company. This company had an abysmal record of making a success of any agricultural undertaking, nevertheless its charter did not expire until 1929 and the Portuguese metropolitan government had little influence until that date.[20]

[17] *ECGR* 6: 3 (July 1929), 266.

[18] See Alan K. Smith, 'Antonio Salazar and the Reversal of Portuguese Colonial Policy', *Journal of African History*, 15: 4 (1974), 662, on radical alterations in colonial policy after Salazar came to power. Although I agree with Smith that the policy changed after 1928, as I argue in this chapter the results were not visible until after 1936.

[19] See Great Britain, Board of Overseas Trade, *Portuguese East Africa*, 1923–33.

[20] Barry Neil-Tomlinson, 'The Nyassa Chartered Company: 1891–1929', *Journal of African History*, 18: 1 (1977), 109–28; and M. Anne Pitcher, 'The Agrarian

Without a commitment from companies or organized supervision from colonial governments, those Africans who were expected to grow the cotton found it an easy crop to switch from and to avoid all together. A report on the Tanganyika territory issued in the *Empire Cotton Growing Review* found that if Africans were left to choose their own living, they did not choose cotton,[21] and this certainly seems to have been the case for Mozambique and Angola. In the years immediately after the legislation was passed, Africans still found more attractive ways of making a living or selected the few remaining options for avoiding work on behalf of the Portuguese. The resulting labour shortages were only exacerbated when Portugal concluded an agreement to increase the export of Mozambican workers for the gold mines of South Africa in 1928. The agreement further reduced labour supplies south of the 22nd parallel where at least one cotton company was attempting to produce cotton—the Anglo-East African Cotton Company.[22]

Just as obstacles to the development of cotton production existed in the colonies, also in the metropole the government did not yet possess the strength to force the acquisition of colonial cotton by the metropolitan textile industry. Due to the uncertain domestic situation and other political and economic demands, the government had not consolidated its power and the disinterest of the industry was too great for the government to override. When international prices rose in the past, manufacturers shrieked for government-sponsored colonial cotton schemes because they were hoping to take advantage of lower colonial prices and duties during phases of rapidly changing international prices. But now that the international price had dropped and stabilized, textile industrialists lost enthusiasm for schemes to develop cotton in Portuguese Africa. Instead, as demonstrated in Chapter 2, the interests of manufacturers at this time lay much more with markets and they directed their lobbying efforts to this concern.

To obtain cotton, manufacturers preferred the international market. Contacts had already been established with international

Question in Mozambique: Recalling the Past to Give Meaning to the Present', M.Phil. thesis (Oxford, 1985), 68–9.

[21] A. J. Kirby, 'The Progress of Cotton Growing in the Tanganyika Territory', *ECGR* 7: 3 (July 1930), 175.

[22] Ruth First, *Black Gold: The Mozambican Miner, Proletarian and Peasant* (Sussex: Harvester Press, 1983), 215–17.

suppliers. Industrialists preferred to maintain a relationship in which quantity and quality were known rather than switch to a system which was still disorganized. The inertia on the part of the industry equally contributed to the failure of the cotton regime in the initial years.[23]

THE DEPRESSION AND INCREASED COLONIAL ADMINISTRATION

Although world prices were appallingly low, during the Depression, cotton production in Mozambique and Angola began to improve, due to greater governmental administration and interest, and reduced options for Africans and companies. The creation of a fixed price and guaranteed placement in the metropole helped provide the necessary incentives to engage in cotton growing. We shall consider the effects of general policies of reorganization in the colonies and then consider the price offered for colonial cotton.

Legislation indicates a gradual strengthening and centralization of metropolitan control during the period. Several months after the 1926 coup, the military government issued a decree remodelling the administration of the colonies. Decree 12,421 of 1926, a precursor of the 1930 Acto Colonial, asserted the political unity of the colonies with the metropole, yet sought the continuation of the old regime of financial and administrative autonomy, 'with more efficient superintendence and taxation by the metropole'.[24] Moreover, it clearly defined and delineated the legislative and executive duties belonging respectively to the metropole and the colonies, and offered a 50 per cent reduction in duties to all metropolitan products exported to the colonies. The decree was, in fact, heavily weighted in favour of metropolitan control and centralization: powers were concentrated in the hands of the Minister of the Colonies. For example, the Colonial Minister was allowed to alter colonial budgets if the colony showed a deficit; he also dealt with all business relating to foreign commerce or treaties. Colonial governors were not allowed to interfere in matters of policy; their

[23] See *OTN* 1926–36: whereas the textile industry frequently lobbied the government to find markets, it hardly ever referred to problems with the cotton supply except in the years when prices were high, as we have mentioned.

[24] Portugal, *Anuário Colonial*, 1927–9, 83.

role was confined to mere administrative tasks such as allocating labour or collecting taxes.

In 1928 the military government followed the decree on administrative organization with a Codigo do Trabalho Indígena (Labour Code). Prior to the Labour Code, private concessionary companies were legally allowed to recruit Africans for work on sugar and tea plantations or in diamond and coal mines.[25] This had frequently resulted in clashes between companies as they struggled for scarce resources. The 1928 Labour Code abolished recruitment by private companies and gave the duty of overseeing labour recruitment to colonial governments. Secondly, the Labour Code stipulated a number of housing and nutritional requirements for Africans to be fulfilled by companies, thus drawing attention to the fact that companies had made little provision for their African workers in the past. Thirdly, the Code stated that Africans had a moral but not a legal obligation to work. While such a stipulation appears to contradict the compulsory element of the 1926 cotton decree, in fact, the new Code gave colonial governments the right to use labour as a form of punishment for criminals, vagrants, and tax defaulters or to fulfil a 'public need', be it building roads, aiding in natural calamities, or meeting a temporary labour shortage.[26] In effect, the state now had legal control over the labour supply and could rely on the loophole to employ labourers as it saw fit.

Under the Salazar Government, the 1930 Acto Colónial (Colonial Act), the 1933 Carta Orgânica (Organic Charter) and Reforma Administrativa Ultramarina (Overseas Administrative Reform Act) incorporated and extended the general principles of the two earlier Acts. Later incorporated into the 1933 Constitution, the Acto Colónial stated that 'the component parts of the Portuguese Colonial Empire stand together in solidarity with each other and with the Mother Country'.[27] The Act reserved to the State all public powers, from levying taxes to the granting of land and mine-prospecting areas. It abolished a type of concessionary system in

[25] See Pitcher, 'The Agrarian Question', ch. 3, on clashes over labour in Mozambique; and Clarence-Smith, *The third Portuguese empire*, ch. 5, for references to the labour struggle in Angola.

[26] Leroy Vail and Landeg White, *Capitalism and Colonialism in Mozambique* (London: Heinemann, 1980), 249–53. The code was altered slightly in 1930 to fit circumstances in Mozambique and it is from this code that we take our discussion.

[27] Portugal, SPN, *Political Constitution*, 'Colonial Act', Part I, Art. 5.

operation since the nineteenth century whereby companies assumed public as well as private functions in a given area including the passage of legislation, the collection of taxes, and the maintenance of a police force. However, no charters were cancelled before their expiration date, they were simply not renewed once they expired. Following expiration, those companies which operated under this system either abandoned their concessions completely or reverted to concessions with economic functions only like those in the cotton areas. For example, the charter of the Niassa Company expired in 1929 and the activities of the company promptly ceased. The charter of the Zambezia Company also expired in 1930 and while its economic privileges remained, its governmental responsibilities reverted to the colonial regime.[28] Although the Mozambique Company's charter did not expire until 1942, by the mid-1930s with the *Estado Novo* having consolidated its power, the company co-operated with the colonial government regarding cotton production.[29]

As in the metropole, by the mid-1930s Salazar's consolidation of power and the demands generated by the Depression gave force to legislation providing for state intervention and reorganization at home and in the colonies. Part III, Article 26 of the Acto Colónial the 'Political and Administrative System', declared:

> Administrative decentralization and financial autonomy are guaranteed to the colonies so far as these are compatible with the Constitution, their stage of development and their own resources, without prejudice to the provisions of article 47 (which allowed for metropolitan control in the event of 'serious situations' developing in colonial finances).[30]

The legislation supposedly guaranteed decentralization and autonomy but these were never, in reality, sufficiently compatible with the metropole's objectives to ensure some measure of independence to the colonies. As in Decree-Law 12,421 of 1926, the Acto Colónial granted wide political and administrative powers to the Minister for the Colonies. With regard to economic matters, metropolitan control was even more omnipresent. Each colony had

[28] Newitt, *Portugal in Africa*, 88.

[29] See E. K. de Queiroz Ribeiro, *O algodão em Moçambique* (Beira: Govêrno do Território da Campanhia de Moçambique, 1939), 64. According to Circular 7,933 of 20 Nov. 1934, the Mozambique Co. was obliged to abide by legislation relating to cotton production.

[30] Portugal, SPN, *Political Constitution*, 1933, 'Colonial Act', Part III, Art. 26.

its own budget, but always subject to the approval of the Colonial Minister. While the economic systems of each colony were established in line with the colony's own needs as well as the demands of mutual exchanges with neighbouring countries, they were also to be established in accordance with the 'legitimate requirements of the Mother Country and the Portuguese Colonial Empire'.[31] As with the Constitution, the evolution of legislation and subsequent events revealed extensive metropolitan involvement in the colonies.

The 1933 Carta Orgânica and Reforma Administrativa Ultramarina translated colonial principles into practice by effecting total administrative reorganization of the colonies. Administrative systems in Angola and Mozambique roughly followed the model in Portugal with government presence at provincial, district, and council levels. In the colonies, High Commissioners were replaced by Governors-General of each colony and a hierarchy of state representation was established right down to the village level. Urban areas were divided into *concelhos* (councils) and rural areas into *circumscrições* (circumscriptions) or *postos* (posts—the most rural).[32] Accompanying administrative reorganization came an increase in police and military presence, followed by numerous *juntas* or economic boards, as we shall see in the case of cotton. Such control evolved slowly and in response to certain needs, the measures taken in the 1930s were merely a 'first-cut' at the metropolitan–colonial problem.

Greater state administration in the colonies facilitated the government's renewed interest in colonial cotton production following the onset of the Depression. With the loss of emigrants' remittances and the quantity and value of Portuguese exports decreasing, the government increased its efforts to reduce foreign imports in order to alleviate the negative payments situation. As in 1926 cotton was an obvious target because it was an import which colonial production could replace and it still accounted for 5 per cent of the value of imports in 1933. Moreover, the rapidly falling international price also conflicted with the desire of the government and the textile industry for stability. From 10$47 escudos per kilo in 1929, the world price had dropped to 4$80 escudos per kilo in 1932; the real price change was just as dramatic (Tables A7–A8

[31] Ibid. Part IV, Art. 35.

[32] *Diário do Governo*, Série I, D-L 23,228 and 23,229 (15 Nov. 1933).

and Fig. A2). Having bought their cotton on the world market in advance of production for one price, textile companies were then obliged to sell their manufactures at a lower price because world prices had dropped during the intervening period of production. Allied with depressed levels of consumption, the instability hurt profits. Many companies in the early 1930s were unable to pay dividends to shareholders and more critically, to repair and replace machinery. As one company explained: 'The cost of prime material dropped continuously during the year which led to a progressive depreciation in the price of our stocks and current manufactures'.[33]

A more vigorous colonial cotton policy formed part of the government's answer to a negative balance of payments and price instability in the industry. In 1932 the metropolitan government established a fixed price to be paid to concessionary companies in charge of ginning and exporting cotton. Price had always determined production in the past so the state decided to play the colonial companies at their own game. To provide incentives to the cotton companies, the government set the price of colonial cotton at 8$00 escudos per kilogram (c.i.f.) in Lisbon.[34] At a time when the international price of cotton had dropped dramatically this represented a 43 per cent increase over the price of American middling which was 5$59 escudos per kilogram in 1933. The subsidy took the form of an export premium for cotton companies. Colonial exporters sold cotton to metropolitan importers at the current world price. The DSA of each colony then reimbursed colonial exporters the difference between the set price of 8$00 escudos per kilogram and the prevailing international price at the time of purchase. This was financed both by a 15 per cent tariff on foreign textiles imported into the colonies (which also served the purpose of favouring Portuguese textiles) and by a tax on foreign ginned cotton imported into Portugal. Since the subsidy took the form of a premium paid to exporters by the government, this

[33] Porto, *Relatório*, 1930, no page.

[34] *Diário do Governo*, Série I, D-L 21,226 (22 Apr. 1932) and 22,616 (clarification 2 June 1933). C.i.f. is an abbreviation for cost, insurance, freight, and means that the price quoted not only includes the cost of the commodity, but also the insurance and freight charges added when the commodity is transported to its final destination. In the commodity trade, it is standard practice to specify what is included in the price, i.e. whether it is wholesale, retail, free on board (f.o.b.—the export price without insurance and freight added) or c.i.f.

meant that the Portuguese textile industry itself did not actually pay the higher price.[35]

Cotton growing for the metropole suddenly became more attractive for both companies and the African workforce. Cotton's placement in the metropole and its price were guaranteed, in comparison with other colonial crops whose markets had disappeared or whose prices had dropped. Options were limited for companies and Africans alike and the prospect of a guaranteed price and market could not be ignored. Moreover, the government helped to recruit the labour, using the loophole of the 1928 Labour Code.

In 1934 the Zambezia Company which had previously grown every crop but cotton stated clearly in its annual company report that it was going to encourage cotton production. Because of the interest expressed by the administration and the premium offered on exports, the Company foresaw that cotton production would leave it with 'a small profit'.[36] Unfortunately for the Zambezia Company, the premium was not paid consistently. In 1933 the fixed price was changed to 7$50 escudos and in 1936 and 1937 it was dropped altogether. In these two latter years, metropolitan textile companies purchased colonial cotton at the lower international price and the premiums were not paid (Table A7). The cancellation of the premiums was so abrupt that many companies including the Zambezia Company had already seeded and could not switch to a more profitable crop for the season.[37] The government's keen interest in saving money makes it easy to believe that the announcement of the premium cancellation was purposely calculated to occur after companies had committed themselves to cultivation. Nevertheless, the promise of premiums did attract a fair number of companies to the scheme after 1932. Companies increased output and switched their markets to Portugal.

Such government intervention and protectionism during this period were not peculiarly Portuguese but were part of an international trend on the part of all imperial powers to see in their colonies a solution to the crisis of markets, prices, and foreign exchange during the Depression. The trend took the form of

[35] D-L 21,226 and 22,616.

[36] Companhia da Zambezia, *Relatório, contas e parecer do conselho fiscal*, 1934, no page (hereinafter Zambezia, *Relatório*).

[37] Ibid. 1937, no page.

greater administrative intervention, price controls, quotas, tariffs, and market-sharing arrangements and varied from power to power, country to country.[38] Like the Portuguese, the British had realized state intervention as the *sine qua non* of colonial production. In 1932, the year that Portugal slapped its 15 per cent tariff on foreign textiles going into the colonies, Britain established the Imperial Preference System whereby British goods were given preference in African markets. Additionally, Britain, France, and Belgium implemented measures and increased efforts to use the colonies to fulfil metropolitan needs.

In cotton, similar measures to those of the Portuguese were enacted by the British in Uganda and Tanganyika. There, European and Indian-owned ginneries were given monopolistic rights over the purchase of peasant grown cotton:

> These measures, eliminating competition between ginners, encouraged the growth of pricing pools and ensured the profitability of ginning companies at the cost of lower prices to producers.[39]

However, whereas in the British colonies colonial governments facilitated the profitability of ginning companies by allowing monopolies and pricing pools, in the Portuguese colonies, the metropolitan government itself guaranteed the profitability of concessionary companies by fixing low producer prices. In addition, the prices of cotton exports from the British colonies were not protected from the vagaries of world price trends; by contrast, Portugal's authoritarian government even set the export price to the Portuguese textile industry. Each approach had its drawbacks, but these differences underline the contrast between authoritarian and liberal styles of imperialism: the former unwilling to entrust price levels to intermediaries, either individuals or companies, or to an economic mechanism such as the market; and the latter, willing and able to use intermediaries and the market to achieve its aims.

The government's high fixed prices had hooked the concessionary companies in the Portuguese colonies, but had not simultaneously excited industrialists in the metropole. A 1934 study by the DSA in Mozambique lamented the fact that national industry seemed uninterested in the colonies:

[38] J. Forbes Munro, *Africa and the International Economy, 1800–1960* (London: J. M. Dent and Sons Ltd., 1976), see ch. 6.

[39] Ibid. 162.

The national factories of cloth, which annually continue to import thousands of contos of cotton fibre from abroad for their use, show a deplorable lack of interest in the development of colonial cotton production according to the model established in D-L 11,994: 'cultivation and sale by *indigenas* (Africans) and purchase, preparation, and export by *colonos* (Europeans, i.e. cotton concessionary companies)'.[40]

The *Boletim Geral das Colónias* reiterated the view of the DSA in 1934: 'In no country has the cotton industry more reason to be interested in the cultivation of colonial cotton than in Portugal. It seems that no country is interested less!'[41]

This is overstated, for the industry's behaviour accorded with its past practice. For reasons cited above, the industry was still ambivalent about purchasing colonial cotton. It meant switching from long-established arrangements in which quality and quantity were more or less known. This still was not the case in the colonies. There, the quality and quantity of harvests could not yet be guaranteed.

Legislation and newspaper reports exposing difficulties with the scheme help to explain why the industry hesitated to embrace colonial cotton. In 1933 additional legislation offered bonuses on top of the premium to try and improve the quality of colonial cotton. Moreover, while the fixed price for cotton exported from the colonies by cotton concessionary companies may have been higher than the international price, the price paid to producers apparently was not, which led to cheating. In 1934 the Governor of Angola reported that in spite of low world prices, Africans in the Catete region were managing to obtain better prices through clandestine trade than the official ones offered by cotton concessionary companies. Having found buyers prepared to flout the law, Africans were accepting the free seed doled out to them by concessionary companies and then trading the crop outside official markets.[42] In fact, different prices appeared to be in operation not only at the producer level but also at the concessionary level in the mid-1930s. *O Comércio* of Luanda reported that Cotonang had even managed to sell its ginned cotton at a price higher than the fixed price of 7$50 escudos because, it was claimed, Cotonang's cotton

[40] Mozambique, DSA, *Algodão*, 10.
[41] *BGC* 10: 106 (Apr. 1934), 111.
[42] *BGC* 10: 108 (June 1934), 189.

crop was of a higher quality and cleaner than that grown in the rest of Angola.[43]

With the differences in quantity and quality of colonial cotton, it was not surprising that the domestic textile industry was reluctant to switch to the colonies. Despite the upsurge in production, the new controls had not been thoroughly and effectively implemented in the colonies. As we shall see, quantity, price, and quality complaints by the Portuguese textile industry were numerous throughout the existence of the colonial cotton regime.

More than anything the decision to fix the price signified the government's determination to pursue autarkic policies with or without industrial pressure. Though there continued to be animosity between the colonies and the textile industry as well as squabbles over prices and clandestine trading, the combined actions of state intervention and protectionism did result in an increase in cotton output from 1932 and a shift of exports to the metropolitan market. Whereas imports of colonial cotton had been 831 tons in 1930, they were 1,793 tons in 1932 and 5,346 tons by 1936 (Table A4). Though imports of colonial cotton still only accounted for 26 per cent of total cotton imports by 1936, their dramatic growth indicated a government commitment to foster colonial cotton growing, to assert control over an area that affected Portuguese finances, and to intervene to secure its objectives. We may note here that the government was emboldened to encourage the import of colonial cotton because the industry was weak; the government was not strong enough to enforce total compliance even then. The period from the late 1930s would reveal just how determined the government was to wed colonial cotton and the metropolitan textile industry.

[43] *BGC* 11: 118 (Apr. 1935), 132–3.

4

Intervention and Industrialization: The *Estado Novo*, 1936–1946

INTRODUCTION

Whereas stabilization and recuperation marked the first period of the *Estado Novo*, intervention and industrialization characterized its second period. From 1936 to 1946, the *Estado Novo* gradually moved from arbitrating amongst competing economic élites to giving greater support to Portuguese industry, particularly to financial-industrial conglomerates. Politically, the *Estado Novo* extended its interventionist approach, gradually becoming more and more authoritarian.

Internationally, the period from 1936 to 1946 saw the occurrence of two wars: first, the Spanish Civil War, and then, World War II, both of which indirectly involved and profoundly influenced Portugal, despite its official neutrality. The two wars reinforced the greater emphasis on industrialization enabling Portuguese industry to expand and allowing financial-industrial conglomerates to strengthen their influence in the Portuguese economy. Aided by increased trade and inflated prices during the Spanish Civil War and World War II, Portugal experienced a period of export-led growth which saw substantial profits made for certain sectors of Portuguese capital. Moreover, World War II also forced Portugal to reduce its imports and to develop industries to substitute for these. With the changing balance in Portuguese business, the *Estado Novo* ceased to play its previous role as arbiter between small and large capitalists, agriculture and industry.

Politically, the two wars also allowed the *Estado Novo* to further consolidate its intervention into Portuguese society. The victory of Franco and his nationalist forces in the Spanish Civil War buttressed the authoritarian-nationalist edifice being constructed by the Portuguese government. It was further supported by the rise of Mussolini and Hitler. Together with the institutional constructions and legal creations of the *Estado Novo*, the international environment contributed to the powerful political position that the regime attained during this period.

THE INSTITUTIONALIZATION OF STATE INTERVENTION

During the period from 1936 to 1946 international trends towards intervention and authoritarianism facilitated Salazar's institutional and ideological attempts to construct a strong state. On its own doorstep, Portugal adopted a public position of non-intervention at the start of the Spanish Civil War in 1936, but there was little doubt that Salazar supported Franco, and Portugal provided port and railway facilities, money, arms, and even volunteers to the rebels.[1] In a speech on 10 September 1936, Salazar vehemently denied accusations by foreign papers of Portugal's involvement on the side of the rebels under Franco.[2] In several other speeches, however, he declared his abhorrence of Communism, which he identified with the cause of Spanish Republicans, and his fear for Portugal's future should the 'Communists' win in Spain.[3] Moreover, Salazar broke diplomatic relations with the democratically elected, Republican government of Spain in October 1936 as part of the policy of 'non-intervention'.[4] When an inquiry into the events at Guernica was called for by the British government in May of 1937, Portugal refused to support it on the grounds that the demand for the inquiry originated from the Spanish Republican government. To support it, said Salazar, would show a kind of favouritism to Franco's adversaries at a time when Portugal as well as Great Britain were committed to a policy of non-intervention.[5]

Salazar's speeches indicate that he had a real political interest in a Franco victory.[6] Perceiving the war in terms of 'Communists' versus 'Nationalists', he expressed no desire to share the peninsula

[1] Martins in Woolf (ed.), *European Fascism*, 323.
[2] Salazar, *Discursos*, ii, no. 14, 181–90.
[3] Ibid. ii, see nos. 15, 27, and iii, nos. 1, 2.
[4] Ibid. iii, no. 17, 219–26.
[5] Ibid. ii, no. 25, 289–93.
[6] Ibid. iii, see nos. 14 and 15.

with a government antithetical to his own, perhaps fearing the domestic consequences of a Republican victory in Spain. He was fearful of the propaganda and activity devoted to the Republicans' cause and he suppressed any information favourable to that side. Franco's victory in 1939, then, bolstered Salazar's position. The Iberian peninsula was now governed by two dictators with similar philosophies though there remained the traditional suspicion of the over-mighty neighbour. Salazar's efforts to consolidate the *Estado Novo* now seemed attainable.

As the Spanish Civil War drew to a close, however, events elsewhere in Europe evoked consternation. Though no doubt comforted by dictatorships not unlike his own in Germany and Italy, Salazar's sense of self-preservation counselled him to fear the internal consequences of Portuguese participation in another world war and the expansionist policies of the Axis powers, and not to anticipate the outcome of the approaching war. As early as 1938 he reiterated a determination not to become involved in the growing European conflict.[7] In a speech to the National Assembly on 22 May 1939 to announce Franco's victory, Salazar took the opportunity to discuss the European situation and Portugal's position in more detail:

> Some months ago the Spanish War ended, and with the nationalist victory were defeated some of the causes of general disquiet. But in the rest of Europe states were dismembered, the independence of others was proclaimed, frontiers were altered—involving transfers of vast territories, and nations were incorporated in diverse ways. As a result, the strategic value and potential of some countries are notably altered.[8]

Salazar defined the position of his country. He did not wish for Portugal to become involved in the conflict but nevertheless he stated that Portugal would honour a friendship agreement with Spain and its old alliance with England (without detailing what such a commitment entailed). At the same time, Salazar affirmed his fierce anti-Communist stance thereby signalling his disapproval of any alliance between the Soviet Union and the Allies.[9]

Even after Portugal's neutrality was declared officially on

[7] Ibid. iii, no. 5, 78.
[8] Ibid. iii, no. 11, 138.
[9] Ibid. iii, see no. 11.

1 September 1939,[10] Salazar continued to make anti-Communist speeches, to trade with Germany, and at least tacitly, to support the Axis. In official diplomatic circles, though, Salazar formally honoured the terms of neutrality.[11] The reasons for this are quite straightforward. First, neutrality allowed him to await the outcome, before rallying to the victor. Secondly, Salazar was forever concerned with internal stability, and too overt support for one or the other side might jeopardize the delicate balance within the armed forces and in the country. This line of reasoning appears to have been correct as domestic opposition groups during the early years of the war were in disarray and had difficulty taking a coherent stance against Salazar.[12] Thirdly, there were economic advantages to neutrality. As a neutral country, Portugal could and did cater to both sides and the profits gained from that arrangement were not inconsiderable.

Unquestionably, however, the political beliefs and values which the Salazar regime had in common with Franco's Spain and the Axis powers, coupled with their initial successes and the exigencies of wartime conditions, strengthened the position of the *Estado Novo* in the early years of the conflict. The war gave the Salazar regime the opportunity to repeat over and over again the advantages of a corporatist ideology, to solidify support, to repress the opposition, and to bolster the strength of the Portuguese armed forces. Principles embodied in the Constitution were repeatedly invoked in order to justify and legitimate any actions carried on by the regime. Nationalism, authority, patriotism, history, family, and discipline were themes constantly referred to in the ideological onslaught embarked upon by Salazar throughout the late 1930s and early 1940s. In several of his speeches, Salazar reiterated:

> We have to create a political structure which is completely Portuguese . . . The principal source of inspiration for the major tenets of our political creation has been our history, traditions, and temperament; in summary—Portuguese reality.[13]

[10] Ibid. iii, see no. 14.

[11] For a discussion of Portugal's role during World War II, see W. N. Medlicott, *The Economic Blockade*, i (London: HMSO and Longmans, Green & Co., 1952), ch. 15; and ii (1959), ch. 11.

[12] Robinson, *Contemporary Portugal*, 67.

[13] Salazar, *Discursos*, iii, no. 4, 31; and see also no. 27, 307–18; no. 28, 319–52; and v, no. 6, 79–88.

These themes were also propagated by the state-created party, the União Nacional (National Union) and its youth organization, the Mocidade Portuguesa (Portuguese Youth).

In its implementation, interpretation, and evolution from the mid-1930s onwards, the interventionist, authoritarian aspects of the Constitution rather than the corporatist or co-operative aspects were used to greater political advantage by the regime. What had appeared to be constitutional concessions to Republicans or workers in 1933 were later manipulated to the state's advantage. Elections were, in effect, strictly controlled and never free. Workers' organizations were used to repress worker demands. Industrial interests were often overridden by those of the government.

Over the course of time the regime sought to depolarize Portuguese society by forcible political demobilization. Independent associations were discouraged or repressed and political organization strictly controlled by the state especially as the regime matured. State employees were required to swear an oath of loyalty to the *Estado Novo*.[14] The middle of the 1930s saw the creation of Legião Portuguesa (Portuguese Legion), the para-military organization, and the modernization of the repressive apparatus of the state. Censorship was enforced; and relaxed only briefly during elections. The government also applied strict controls to the founding of new publications from 1936, leaving it to the censorship services to decide on the financial viability of any new publication and the moral and intellectual suitability of its directors.[15] The years from 1936 to 1939 also saw the greatest number of political prisoners taken by the regime. From a high of around 1,400 in 1936, the number fell to approximately 100 prisoners in 1940 as the regime devised more subtle means of control. Subsequent years brought periodic increases in the number of political prisoners especially during election periods.

During the second decade of the authoritarian regime, the government continued to consolidate and legitimate its political and economic power through the creation of institutions. *Sindicatos*, *grémios*, *casas do povo*, and *casas dos pescadores*, continued to be founded: by 1950, there were 312 *sindicatos* and 498 *casas do*

[14] Martins in Woolf (ed.), *European Fascism*, 322–9.
[15] Cruz, *Monárquicos e republicanos*, 81.

povo.[16] In 1936 the creation of organisms of economic co-ordination increased state intervention in key sectors of the economy. During World War II, the state used its power of intervention to maintain low wages and salaries, and to enforce long working hours. Added to the shortages of food and other necessities, these hardships prompted strikes throughout Portugal from 1942 and the formation of an official opposition group in 1943.[17] This increase in political activity coincided with the weakening of the Axis position in the war. With the Allied victory in 1945, the Salazar regime witnessed the end of its unchallenged dominance of Portuguese society.

THE *ESTADO NOVO* AND THE ECONOMY

Financially, the late 1930s and 1940s were active years for the Portuguese government. Aided by increased demands for Portuguese products brought by the Spanish Civil War and World War II, as well as his own careful budgeting and balancing, Salazar placed Portugal on sound financial footing in these years. Due to import reductions, the acquisition of new markets, the improvement of trade with the colonies, the development of new exports (mainly in tin and wolfram exported to the Axis) and the increase in export values caused by the war, Portugal's chronic trade deficit was eliminated in the years 1941 to 1943.[18] Although negative trade balances recurred from 1944, revenues from services, emigrants' remittances, returns on capital invested abroad, and the expenses of refugees, diplomatic missions, and travellers in Portugal resulted in a positive balance of payments of more than 17 million contos by 1946. Portugal had a healthy foreign exchange and gold reserves which increased from 1.7 million contos in 1939 to 18.5 million contos in 1946.[19] During the 1940s, the escudo remained stable at about 101 escudos to the pound (Table A1), though the period was one of acute shortages for much of the population.

Moreover, the state engaged heavily in public works and infrastructure. 'Public works became a sort of *leitmotiv*, a subject that revealed the new Portugal on the move, and indeed one of the great

[16] Schmitter, *Corporatism and Public Policy*, 17 and 20.
[17] Rosas, *Portugal entre a paz e a guerra*, 391.
[18] Ibid. 141.
[19] Ibid. 155.

achievements of the Salazar period.'[20] The interest in public works began slowly. A fifteen-year Lei de Reconstituição Económica (Law of Economic Reconstruction), decreed in 1935, allocated 6.5 million contos for defence, ports, roads, railways, irrigation, hydroelectric works, and communications.[21] Between 1936 and 1939, 2.2 million contos were spent, but approximately half went to defence; transport, agriculture, and public buildings received most of the remainder.[22] With the economic opportunities brought by World War II, however, expenditure on public works and development increased. Production of electricity increased from 425,729 kilowatt/hours in 1938 to 545,756 kilowatt/hours in 1945, though this amount still insufficiently satisfied demand.[23] Railways were modernized; the road network, which was only 13,000 kilometres in 1926 doubled by 1950.[24]

In the shadow of a major world war, the state was able to accomplish other economic goals too. For example, the period saw a reinforcement of links with the colonies helping to save on foreign exchange and bolstering the state's ideological emphasis on the role of empire and the benefits of nationalism. Trade with the colonies saw notable increases between 1939 and 1946. Whereas 13 per cent of Portuguese exports went to the colonies in 1939, by 1946 Portugal was exporting 23 per cent of its products to its overseas possessions. Imports from the colonies showed a similar increase rising from 11 per cent in 1939 to 15 per cent by 1946.[25]

The war also allowed the regime to proceed with its protectionist and nationalistic plans for industry away from international scrutiny. Not only did the domestic market become dominated almost entirely by national industry but the 1943 Lei de Nacionalização de Capitais (Law of Capital Nationalization) reserved key areas of the Portuguese economy for national investment. In addition, foreign investors who already owned businesses in Portugal and the colonies were encouraged to reduce their share to 40 per cent or below, though in the post-war period this provision was dropped in the colonies and only sporadically applied in the metropole.[26] Such

[20] Marques, *History of Portugal*, ii. 196.
[21] *Diário do Governo*, Série I, no. 118, Law 1,914.
[22] Rosas, *Nos anos trinta*, 201–2.
[23] Rosas, *Portugal entre a paz e a guerra*, 112.
[24] Robinson, *Contemporary Portugal*, 137.
[25] Rosas, *Portugal entre a paz e a guerra*, table IV.
[26] Ibid. 213.

measures enabled Portuguese industrialists to make huge profits out of the war, while the regime extended its nationalist objectives.

Alongside its efforts to improve the economy and to benefit industry, the state extended its grip on it. In July 1936, a month before the start of the Spanish Civil War, the government passed Decree-Law 26,757 creating the organizations of economic co-ordination. These were a series of commissions, *juntas*, and institutes which formed an institutional and bureaucratic link between the government and economic groups. They were responsible for regulating imports and exports, fixing prices, limiting supply, determining product quality, and settling wages. Commissions were set up to regulate imports and exports into and from the metropole and the colonies. For example, the Comissão Reguladora do Comércio de Algodão em Rama (CRCAR) composed of government officials, textile industrialists, cotton importers, and exporters dealt with all aspects of the cotton supply once it reached Portugal and we shall discuss it in detail in the next chapter. *Juntas*, on the other hand, were responsible for coping with colonial or metropolitan exports. For example, the Junta de Exportação de Algodão Colonial (JEAC) organized cotton exports from the colonies and the Junta Nacional de Exportação das Frutas (National Board of Fruit Export), the export of fruit from Portugal. The regime created institutes mainly in those sectors which produced for export, already had *grémios*, federations, or unions, and required 'official guarantee of their quality and category' such as wine or cereals.[27]

World War II allowed yet greater government intervention in the economy because the government had to regulate salaries, control imports and exports, and fix prices in order to maintain its goal of economic stability and to satisfy British concerns about its neutrality. In 1939 Decree-Laws 29,904 and 30,137 allowed the government to regulate imports and exports as it saw fit. Further decrees created regulatory commissions for iron and steel, coal, pharmaceutical products, and vegetable oils in order to monitor the imports of these materials with the advent of the war.[28] Moreover, through the Conselho Técnico e Corporativo do Comércio e

[27] D-L 26,757 quoted in Manuel de Lucena, 'Sobre a evolução dos organismos de coordenação económica ligados à lavoura', Part I, *Análise Social*, 14: 56 (1978–4), 840.

[28] Lucena, 'Sobre', Part II, *Análise Social*, 15: 57 (1979–1), 134–5.

Indústria (Technical and Corporative Council of Commerce and Industry) created within the Ministry of Commerce and Industry in 1936, the regime increased its control over industry. The Technical Council supervised the regulatory commissions, studied Portugal's trade, and co-ordinated the activities of all economic organizations. These powers gradually expanded in the 1940s: in 1944 decisions of the commissions of economic co-ordination were subject to the approval of the Technical Council under the terms of Decree-Law 34,329.[29] Even the decisions of voluntary *grémios* were subject to the approval of the Conselho Corporativo (Corporative Council) a government body consisting of Salazar, two professors of Law from Lisbon and Coimbra Universities, and the ministers of Justice, Public Works and Communications, Commerce and Industry, and Agriculture.[30]

The organizations of economic co-ordination were nominally pre-corporative. They were promoted as intermediary bodies which would disappear with the creation of self-regulating corporations.[31] However, their duties often extended far beyond the ostensible purpose for which they had been created. Secondly, their actions rarely represented the views of the sectors they controlled. Thirdly, their existence only served to increase state involvement, notably in important sectors of exports or domestic consumption such as fruit, wine, textiles, cork, and wheat production, or in those of strategic value for the economy such as coal, iron, and steel. Finally, the organizations of economic co-ordination aided the consolidation of powerful economic groups within these sectors.[32] Even when the corporations themselves were belatedly set up in the late 1950s, these bodies did not disappear.

Supposedly the autonomous instruments of a self-directed economy, corporative organizations became the means by which the government intervened to a greater and greater extent in the economy.[33] The notion of self-direction propagated by the Constitution was a myth. Any corporatism that existed was a kind of corporatism by default in which an interest group wrangled con-

[29] Ibid.
[30] Ibid. 136, see n. 64.
[31] Ibid. Part I, 838.
[32] See esp. Part II of Lucena's study on the organisms of economic co-ordination created in the agricultural sector.
[33] Lucena, 'Sobre', Part II, 118.

cessions out of the government by using threats of political or economic sabotage. Far from establishing a self-directed economy, the *Estado Novo* created a hierarchical, centralized, and statist one. Many *grémios* and/or industrial sectors were subject to economic co-ordination organizations which in turn were subordinate to decisions taken by the Conselho Técnico within the Ministry. These organizations of economic co-ordination had excessive power and made arbitrary regulations. They were often corrupt and showed favouritism especially to larger groups; and they were excessively bureaucratized.[34] We will investigate the role of one of them, the Cotton Regulatory Commission (CRCAR), in more detail in Chapter 5.

The second area of state intervention was the Industrial Regulation law. Initially passed as a temporary measure during the crisis years of the early 1930s, by 1937 it was a key instrument of policy: designed to protect or create monopolies, shield traditional and emerging economic groups from competition, and preserve and maintain government supervision over the financial activities of industrial, commercial, and agricultural enterprises. Law 1,956 of 1937 was a bold restatement of the 1931 law. It broadened the branches of industry whose activities were subject to government authorization. It is no coincidence that this law was passed in 1937, one year after the start of the Spanish Civil War. Profiting from the increased sale of goods to Spain, Portuguese industry clamoured to meet the demand. Idle plant was re-employed bringing excess capacity and cutthroat competition. Deputy Ruy Ulrich described the nature of the growing competition in Portugal when the Bill was presented to the Corporative Chamber in 1937:

> If an industrialist creates a new industry which gives him good profits, justly earned through his entrepreneurial effort and innovative courage in the face of risk, soon one, two, ten competitors emerge causing difficulties to a business which would have remained profitable for a few . . . It is a deplorable feature of our national behaviour, but until these habits are changed, the state must correct these excesses by legal limitations.[35]

Law 1,956 sought to expand the number of industrial sectors requiring government authorization for the construction of or

[34] Ibid. 140–66.

[35] Ruy Ulrich quoted in Fernando Cruz, 'Política industrial: Algumas reflexões sobre o condicionamento', in *Relatórios de problemas gerais*, II Congresso da Indústria Portuguesa (Lisbon: AIP, 1957), 18.

changes in buildings, machinery, and products. These included: industries which produced over and above the levels of domestic consumption and therefore had production available for export; industries which used expensive foreign equipment; industries which employed numerous personnel or might be tempted by mechanization leading to possible redundancies; industries which made products needed by economically or socially important Portuguese industries; and finally, industries that produced articles principally for export.[36]

Section III of the law established that government authorization was needed for:

1. new establishments or the reopening of a factory shut for two years or more;
2. any modifications which involved major changes in the industrial register (a description of an enterprise's facilities and major functions);
3. the transfer of property to foreigners or to other nationals if a change of locale was involved.

Section VII required that industries subjected to regulation periodically submit to the government the cost of articles for sale, raw material prices, work conditions, and salaries paid.[37]

Up to 1942 further decrees expanded the number of industries subject to regulation and clarified terms used in previous legislation. Of the official exemptions from the regulation, Decree-Law 28,466 of 1938 listed mostly those enterprises with inelastic demand, for example, jewelry repair shops, photographic labs, suitcase-making factories, and tinsmith shops; and, as before, home industries using machinery of under 5 horsepower.[38]

Greater centralization in this period was accompanied by a greater emphasis on industrialization. The failure of the wheat campaign in 1936, the institution of guaranteed prices for agricultural produce during the 1930s, and the resistance by agricultural interests to any land reforms led to stagnation in the agricultural sector and the emphasis shifted to industry.[39] In Chapter 1, we

[36] *Diário do Governo*, Série I, Law 1,956 (17 May 1937).

[37] Ibid.

[38] Mattos, *Condicionamento Industrial*, no page.

[39] Miriam Halpern Pereira, *Assimetrias de crescimento e dependência externa* (Lisbon: Seara Nova, 1974), 38 and n. 31. We will refer again to the stagnation of

noted the increasing power of several family groups as well as the embryonic growth of interlocking financial, commercial, and industrial enterprises prior to the *Estado Novo*. During the 1930s and 1940s, the process continued:

In the shadow of the *Estado Novo* and the instruments of its economic policy . . . were created the bases for the implantation and later growth of nuclei of industrial and financial monopoly capital.[40]

The state's contribution to this process of consolidation in industry extended from direct, blatant use of corporate organizations and Industrial Regulation legislation to simply turning a blind eye to the existence or formation of monopolies (as shown in Ch. 1). In addition to being interventionist, the organizations of economic co-ordination helped to create monopolies or oligopolies in traditional industries and particularly in those which were not highly concentrated such as ceramics, textiles, and glass.[41]

Industrial Regulation legislation was also used to help the consolidation of monopolies in industry. The justification for such an extensive governmental grip on the economy was repeatedly given as the need to protect Portuguese business, to control unorganized competition, and to create the conditions for a healthy economy. Yet even before the entire body of legislation was enacted, the policy met with bitter condemnation—its real purpose obvious to those initially affected. Julio Martins, the general director of the Companhia das Fábricas Cerâmica Lusitânia responded to a survey on Industrial Regulation in 1936 with harsh criticism of the legislation. Rather than developing national production, defending established industries, increasing raw material supplies, protecting industry from foreign competition, and creating harmony amongst industrialists, the regulations had simply encouraged bureaucracy, monopoly, and fraud:

In reality, we see that in the shadow of 'Regulation' barely legal transactions take place which threaten the reputation of public services, even those of the state itself! Protected by the law, groups and cliques formed monopolies in order to trammel any legal and loyal competition. These

agriculture in Chs. 7 and 10, as it was a persistent problem throughout the course of the regime.

[40] Rosas, *Nos anos trinta*, 52.

[41] Lucena, 'Sobre', Part II, 150–1; and Rosas, *Nos anos trinta*, 218.

monopolies thwart the initiative of enterprising and active industrialists. As a consequence, if they do not possess the strength to fight the monopolies, these industrialists either resort to illegal practices, allow their activities to stagnate, find loopholes in the law, or wait for centuries for the simplest authorizations or licences; while their competitors ingenuously transform themselves by somersaulting over reams of stamped paper.[42]

Martins himself prudently dissociated the state from the acts and abuses of 'groups' and 'cliques', but it is clear what is going on here. Even as it continued to keep wages and salaries low, and to protect industry and agriculture generally, the regime was also aiding and abetting the process of concentration in ceramics and other industries through the use of Industrial Regulation. There was also a *grémio* created for ceramics in 1938. In the next chapter, we shall examine the process in greater detail for the textile industry but similar developments were also occurring in glass, cereals, and resin. Additionally, where industries subject to Industrial Regulation required substantial investments in technology or plant, the larger groups, wealthier families in Portugal, or even foreign capital were allowed to move in, despite Industrial Regulation and the presence of corporate bodies. Though these larger groups did not always eliminate smaller firms they could effectively control the market. This is evident in the sectors of inorganic chemicals, paper, sugar, metallurgy, and soap.[43]

On the other hand, informal patronage or simply turning a blind eye seemed to have been the government's norm for many of the newer industries. In the sector of intermediate goods and equipment, concentration was effected by the prevention of competition or by government concessions granted exclusively to one enterprise.[44] The cement monopoly in 1934 and the superphosphate monopoly built up by CUF in the 1930s are examples of the former phenomenon; petroleum-refining, naval construction, and electrical energy, examples of the latter. In the case of petroleum-refining,

[42] Julio Martins, 'Estudo enviou a Comissão de Estudo da Reforma do Condicionamento Industrial', Separata da *Revista Arquitectura Portuguesa e Cerâmica e Edificação* (30 Dec. 1936). Mr Martins was probably referring specifically to the takeover of another ceramics company, Carvalinho by the Fábrica de Louça de Sacavém in the 1930s, making this factory the largest in the sector. Otherwise the ceramics industry was characterized by a high level of dispersion at the time. See Rosas, *Nos anos trinta*, 219–20.

[43] Ibid. 248–52. Rosas included textiles in the group of industries which wealthy families moved into in the late 1930s but I found no evidence of this.

[44] Ibid. 248.

the government even participated, providing one-third of the capital in the creation of SACOR in 1938 along with Espírito Santo and various other shareholders.[45] For the three industries of petroleum-refining, naval construction, and electrical energy, Industrial Regulation was used essentially to defend favoured firms from competition; within their sectors they could do as they liked.

In the sector of consumer goods too, the regime allowed prior monopolies or investment by foreign companies to continue as in the case of tobacco, matches, and beer. Interestingly, in the sectors just mentioned, no corporate bodies or organisms of economic co-ordination ever existed. They were not necessary: concentration had already occurred.[46]

While legislation, corporate institutions, and informal patronage were instrumental in the way that capital accumulated, it was the Spanish Civil War and most importantly, World War II which made possible the accumulation of large amounts of capital and accelerated concentration in industry. Had external events not been propitious, it is unlikely that the regime would have shifted from its role of arbiter. As it was, cloaked in its commitment to neutrality, Portugal substituted for imports by building up many newer industries, reserved its domestic and colonial markets for its own industries, proceeded to trade with friend or foe in the early years of the war, and benefited from increases in export prices due to shortages. Britain repeatedly concluded import and export rationing agreements with Portugal on products such as petroleum, wolfram, coffee, cotton, and other metropolitan and colonial products. Portugal repeatedly broke the agreements to trade with Germany, Italy, Spain, and Britain.[47] Almost to the end of the war, Salazar continued to avoid firm agreements with either side. He may have been making his own political situation in Portugal more secure, but it is patently clear that:

> he had no desire to lose prematurely the real but precarious prosperity of the war years; with considerable subtlety and some genuine passion he maintained the rightness of a neutrality which extracted abundant profits from both sets of belligerents.[48]

[45] Ibid. 215. By the late 1940s, the Banco Português do Atlantico became the third biggest shareholder.

[46] Ibid. 248.

[47] See Medlicott, *The Economic Blockade*, i, ch. 15; and ii, ch. 11, for a discussion of the haggling over rationing agreements between Britain and Portugal.

[48] Ibid. ii. 582.

Although total tonnage exported actually dropped from 1939 to 1946, the value of exports nearly tripled. Even discounting for inflation, the increase is impressive. Portugal exported 1,029,336 contos worth of goods in 1939 and 2,906,276 contos in 1946: traditional exports such as cork, textiles, conserves, and resins saw sustained rises during the period; wolfram and metals, fleeting but dramatic increases. Profits from exports, particularly wolfram continued to be made at least up to June 1944, the date that Portugal prohibited the sale of wolfram to the Axis at the demand of the Allies.[49]

By the end of the war, not only had Portuguese industry prospered but concentration had taken place. With state support, as we have seen, many of the large families either expanded into new areas or consolidated their control of a single sector. The petroleum-refining company, SACOR, was one of the biggest monopolies to emerge from the war; tobacco, iron, and steel metallurgy also centralized. The number of employers halved from 1930 to 1940 and then reduced by another 30 per cent from 1940 to 1950, while the number of employees increased by one-fifth over the two decades as Table 4.1 shows.

The phenomenon repeated itself in the colonies. CUF had extensive interests in oilseeds, copra, palm oil, insurance, and commerce. Champalimaud bought up the Matola cement works in Mozambique in 1944, the year that controlled industrialization was allowed to proceed in the colonies. The newly formed Banco Português do Atlantico financed the first textile mill in Angola from 1944 and later the first mill in Mozambique, as we shall discuss in a forthcoming chapter. Espírito Santo increased and improved its agricultural interests in the areas of sugar and coffee and was active in the Mozambique Company when that concession reverted to a limited company in 1942.[50]

The accumulation of capital, the formation and consolidation of monopolies and oligopolies in important sectors of the economy, the links with the colonies, the improvement of the general financial situation of the country as a whole, and increased industrialization and investment testify to the many changes which occurred during the Spanish Civil War and World War II, but it is important not to

[49] Rosas, *Portugal entre a paz c a guerra*, 148.
[50] Clarence-Smith, *The third Portuguese empire*, 168–9.

TABLE 4.1. *Number of Employers and Employees in Manufacturing, 1930–1950*

Year	Number of Employers	Number of Employees
1930	644,300	494,000
1940	336,600	767,800
1950	214,600	624,700

Source: Rafael *et al.*, *Capitalismo*, 13.

overstate the case. In the countryside as well as the towns, the two wars brought few benefits to the majority and created hardships for many who were deprived of basic necessities such as food, fuel, and clothing. During World War II, the government fixed wages to control inflation. In real terms, a peasant who had earned 35 escudos per day during the harvest in 1942, only earned 16 escudos per day in 1943 even though costs of living were rising (see Fig. A1).[51]

Secondly, wartime conditions allowed great profits to be made but did not result in modernization of traditional export sectors such as textiles, cork, shoes, ceramics, and food and beverages. Certainly, difficulties obtaining new equipment or more advanced technology played a large part, but it is equally true that the reuse of idle plant and the resurgence of cottage industries explain the existence of outdated technology and obsolete machinery. Moreover, easy placement of products in highly protected markets and controlled wages deluded many Portuguese industrialists into making overly optimistic projections about their trading prospects after the war and left many apathetic about modernization. While profits grew on average around 115 per cent from 1939 to 1945, gross fixed capital formation declined by 30 per cent in the manufacturing sector between 1938 and 1945—confirmation of the entry of backward equipment and the lack of reinvestment.[52]

Thirdly, although the state began more openly to favour power-

[51] See Wiarda, *Corporatism and Development*, 167–70, on strikes over food and the rise in the cost of living during World War II; and Rosas, *Portugal entre a paz e a guerra*.

[52] Ibid. 202 and 229.

ful economic interests, it did not wholly abandon other sectors in society. To preserve its bases of support, it continued to offer general protection, though arbitrarily and haphazardly. Fourthly, while the state aided capital accumulation and concentration, it did not anticipate the social and political changes that such action might bring—an oversight for which it paid dearly in the post-war period.

Finally, the regime maintained its own priorities, political as well as financial, and increased its intervention as we have shown. It continued to adhere to the principles of a stable currency and a positive balance of payments. Salazar, ever mindful of the precarious fiscal state of Portugal, was also determined that the state should share in the bonanza. He complained about unscrupulous profit makers and criticized those who were making excess profits from the war—perhaps in order to justify the high taxes he slapped on 'war profits' in 1942.[53] Throughout the period, the regime maintained the Industrial Regulation laws and its corporate organizations enabling it to intervene in most of the country's economic sectors. The trends were now very pronounced: the state was more interventionist, the emphasis was on industry, in whose prosperity the government now had a clear and immediate stake.

In the next chapter we shall trace the changes of the textile industry from 1936 to 1946 detailing its relationship with the regime and its reliance on the colonies. As for the colonies, their relationship *vis-à-vis* the metropolitan economy also altered during the 1940s. As we shall show in Chapter 6 on cotton during World War II, the colonies assumed a vital importance in the supply of raw materials to Portugal during the war.

[53] Law 1,989 stated that if profits were over 20% of normal return, then they were subject to progressive taxes. A normal return was calculated on the basis of returns for years 1937 to 1939 or failing those, 1941. Taxes varied from 20% to 50% according to the amount of profits. The law was applied to textiles, sugar, coal, cement, cork, machinery, paper, and resins and perhaps more. See *Bol. da DGI*, 5: 236 (18 Mar. 1942).

5

The Textile Industry at War: New Markets, Old Machinery

From the mid-1930s, government policies, the consolidation of power by the *Estado Novo*, and external developments, first the Spanish Civil War and then World War II, combined to bring greater government control and new export opportunities for the textile industry. Central aspects of government control included ensuring the supply of colonial cotton to the metropole, reserving colonial markets for the industry, and regulating structural expansion within it.

By the early 1940s, state intervention through Industrial Regulation and the Cotton Regulatory Commission, together with international events, had managed not only to tie the textile industry to colonial cotton, but also to control its growth and to secure markets for the industry. Shortages of foreign cotton forced the industry to rely on colonial cotton imports and the persistence of domestic competition facilitated the government's control over expansion. For many in the industry, however, increased export opportunities diverted their attention away from the negative consequences of these developments. The war brought financial prosperity to many; equally, its closure revealed the industry's dilapidated state and the extent of government intervention.

THE CAPTURE OF NEW MARKETS

For the textile industry, the 1930s and 1940s were a time of expanding markets, internally and externally, as they were for most Portuguese industries. After 1936, due to state protectionist policies and wartime conditions, the metropolitan share of the colonial market increased. In 1939, due to preferential customs duties and protectionist policies, a total of 3,251 tons of Portuguese textiles were exported to the colonies, as compared with a total of

2,950 tons in 1935 (see Table A5). The amount probably would have been higher were it not for competition from Japanese textiles which began to pour into Mozambique. Portugal's Asian colony of Macau had been left untouched by customs revisions in the early 1930s and as a result, goods from other countries reached Angola and Mozambique via Macau.[1] In Mozambique alone (excluding the area of the Mozambique Company), whereas the metropolitan share of textile imports had been 55 per cent in 1935, by 1937 only 34 per cent of textile imports were from Portugal due to Japanese competition there.

Japanese textiles were quoted as being 30 to 50 per cent cheaper than Portuguese textiles and their appearance gave the metropolitan industry plenty to complain about.[2] In 1936 and 1937 the Companhia de Fiação e Tecidos de Fafe referred to Japanese textiles going to Mozambique through Macau and thereby competing with Portuguese textiles.[3] Fortunately, exports to Spain from 1936 to 1938 temporarily relieved the export problem. Discussing the two situations, the company argued:

> Competition and the contraction of colonial markets, owing in part to contraband Japanese articles going through our colony of Macau, resulted in a deplorable stagnation of business this year which the temporary export of some manufactures to neighbouring Spain only slightly managed to diminish.[4]

In late 1936 the government passed new protective tariffs which applied to all of the empire 'in order to combat foreign competition and especially that of Japan which, like a cyclone, threatens to destroy the very foundations of all the nations' economies'.[5] However, Japanese imports continued to threaten imports from Portugal until World War II. *Indústria Têxtil*, the journal of the industry, noted Japanese competition in the colonies in 1939 even suggesting that textile industries be established in the colonies to counteract the competition.[6] Such an idea had already aroused fierce protest

[1] Clarence-Smith, *The third Portuguese empire*, 159.
[2] *OTN* 16: 189 (1935), 2499.
[3] Companhia de Fiação e Tecidos de Fafe, *Relatório e contas da direcção e parecer do conselho fiscal*, 1935, 1937 (hereinafter Fafe, *Relatório*).
[4] Ibid. 1937, 6.
[5] *Indústria Têxtil*, 1: 5 (Apr.–May 1937), 13.
[6] *Indústria Têxtil*, 3: 22 (Mar.–June 1939), 79–80.

from the industry as seen in the Amboim example but would be reviewed during the 1940s.

The advent of World War II saw the elimination of Japanese textile competition in Mozambique and Portuguese textiles finally captured that market. A greatly relieved industry praised governmental efforts to ensure the export of national textiles to the colonies:

> Let us hope that the apprehensions and uncertainties in which our industry has lived recently have disappeared as a result of the measures adopted by the government for the placement of metropolitan textiles in our colonies.[7]

Total exports to the colonies equalled 7,518 tons by 1946 (Table A5).

The destruction of the Catalan textile industry occasioned by the Spanish Civil War additionally afforded the Portuguese textile industry the opportunity to supply that market until 1939 as cited above. From only 19 tons in 1935, exports abroad rose sharply to 187 tons in 1936. In 1937 and 1938 exports continued to rise to 2,517 tons and 3,114 tons respectively. These are official figures and do not account for the clandestine trade which occurred before and during the conflict.

With World War II following on the heels of the Spanish Civil War, exports increased and prices shot up as well. The disruption of trade patterns, and cotton and textile shortages led to increased demand for textiles plus steep rises in their value. Neutral Portugal benefited greatly. France and Morocco bought Portuguese textiles and other countries also received exports from Portugal as they abandoned or were abandoned by, traditional suppliers. Some exports may also be unaccounted for, especially those included in the substantial contraband trade with Germany in defiance of British regulations. Officially, Portugal experienced large if erratic jumps in textile exports at this time, undoubtedly due to the economic blockade periodically practised by Britain, but these jumps were compensated for by price rises. For example, in 1941 exports abroad equalled 4,085 tons, yet in 1945 only 1,662 tons; in 1944 and 1946 they were only 708 and 428 tons respectively (Table A5). However, the price per unit ton of textiles exported rose 253 per cent between 1939 and 1945—more than the rate of inflation.[8]

[7] Alcobaça, *Relatório*, 1939, no page.
[8] Rosas, *Portugal entre a paz e a guerra*, 122.

State protection followed by the war also led to a drop in foreign competition in the domestic market. In 1936, 70 per cent of imports had been replaced by national brands. By the late 1930s, Portuguese textiles had replaced almost all foreign textile imports. However, the attractiveness of the export market negatively affected the domestic one. Many textile manufacturers interested in making the largest profit possible abandoned the domestic market—preferring to sell in more lucrative colonial or international markets. Alternatively, calculating that consumers had no choice but to purchase whatever items were available, many manufacturers stopped supplying the popular end of the domestic market and came increasingly to supply the market with high quality, expensive cotton goods. Two results followed. As shortages occurred domestically, a black market arose. Concomitantly, many cottage industries, which had been living a hand to mouth existence since the Depression, re-employed old machinery and replaced the products abandoned by larger textile manufacturers. In 1942 the government published an extensive list specifying the cost and type of items manufacturers were required to produce for the domestic market. Twenty-five per cent of production was affected by the new legislation; by 1944, 90 per cent of production was subject to government regulations. Nevertheless a black market flourished and cottage industries proliferated.[9]

Ever ready to turn a situation to its own advantage, the government filled its own coffers by slapping huge taxes on 'war profits' in 1942. War profits were considered to be any profit over and above what was normally expected from sales plus 20 per cent. Normal profits were understood as the average profits realized in 1937, 1938, and 1939.[10] In their reports, companies wavered between bragging about profits and avoiding huge taxes. One company reported very good results and reasonable profits and then tried to deny that these were 'war profits'.[11] There was some truth to the claim, for profits had been abnormally low following the Depression while purchases of machinery and raw materials were now indefinitely delayed, but the government was unsympathetic to these pleas.

In spite of peaks and dips in the quantity of textiles exported

[9] Ibid. 122–3.
[10] *Diario do Governo*, D-L 33,582 (23 Mar. 1944) and 1,989 (6 Mar. 1942).
[11] Guimarães, *Relatório*, 1941, 6.

abroad, the confusion in the domestic market, and taxes on 'war profits', companies reported sizeable profits during World War II—some as much as 100 per cent.[12] One company reported vigorous sales throughout the war after speaking of a crisis in its factory in 1938 and 1939.[13] Other companies also reported good sales even in years when the official quantity of reported sales dropped, illustrating that values had increased or that contraband trading was going on.[14] Towards the end of the war, these profits were also boosted because colonial cotton had replaced almost all of foreign cotton at a price which was 22 per cent lower than the world price by 1945. By the final year of the war, reserve accounts had increased 315 per cent since 1939 and total profits were 25,671 contos—twice the total capital of firms in the industry.[15]

THE CREATION OF THE COTTON REGULATORY COMMISSION

Buoyant sales domestically, colonially, and internationally brought by protectionism and war blinded industrialists to the tentacular control that the regime extended over all aspects of the industry through government organizations and legislation. Determined to link the textile industry to colonial cotton production and to monitor the industry's activities, the government created the Comissão Reguladora do Comércio de Algodão em Rama (CRCAR). Justifying the need for such a regulatory commission, the government said:

> As with the already existing regulatory commissions, the same policy of defending our commercial balance will be followed with cotton imports from abroad. We will give preference to those countries which usually buy our products. On the other hand, the duty of the new regulatory commission will be to co-operate with similar organisms in the colonies, to stabilize the colonial cotton trade, and to ensure its placement in the metropole.[16]

[12] Carlos Bastos, *Indústria e arte têxtil* (Porto: Tip. Portugalia, 1960), 280.

[13] Guimarães, *Relatório*, 1941, 6.

[14] See e.g. Companhia Fabril do Cávado, *Relatório*, 1942–4 (hereinafter Cávado, *Relatório*); and Alcobaça, *Relatório*, 1942–4.

[15] Rosas, *Portugal entre a paz e a guerra*, 123.

[16] *Bol. da DGI*, 1: 19 (19 Jan. 1938), D-L 27,702 (15 May 1937).

In addition to guaranteeing the placement of colonial cotton in the metropole, the role of the CRCAR as stated in Decree-Law 27,702 of 1937 was to co-ordinate colonial cotton production and the needs of the national textile industry; regulate the supply of cotton to factories, particularly the foreign cotton supply; and promote the classification of standard types of cotton. The commission was financed by a tax of 10 centavos on each kilogram of cotton imported from abroad and 7 centavos per kilogram of imported colonial cotton.[17]

While much of the CRCAR's work dealt with ensuring the supply of colonial cotton to the textile industry, its purpose was also 'to create corporate consciousness and to develop the feeling of solidarity among all branches of the activities which it disciplines and directs'.[18] In addition to its quite specific task in co-ordinating the cotton supply, therefore, the commission was endowed with wider political and ideological duties which allowed it to intervene with impunity in the affairs of the textile industry.

All importers and cotton mill owners were obliged to subscribe to the CRCAR. Those who refused risked a fine under the decree-law. The commission itself consisted of a government-appointed president and vice-president, a representative each from the colonial cotton board, the textile industry, cotton importers, and the Technical and Corporative Council of Commerce and Industry.[19] The representative of importers was supposed to be the head of the Grémio Nacional dos Importadores de Algodão em Rama (GNIAR), but since the importers' *grémio* was not formed until 1939, the Minister of Commerce and Industry appointed the importers' representative in 1937. He also appointed the representative for the textile industry, since the textile industry did not have a *grémio* either; and the Technical Corporative Council was in fact a part of the Ministry of Commerce and Industry, responsible for debating the technical merits of applications for new machinery or plant expansions. The Minister for the Colonies appointed a colonial cotton representative until the colonial cotton board—the Junta de Exportação de Algodão em Rama (JEAC), itself government-appointed, was formed in 1938. In effect, this meant

[17] Ibid.

[18] Cotta, *Economic Planning*, 137.

[19] *Bol. da DGI*, 1: 19 (19 Jan. 1938), D-L 27,702 (15 May 1937).

that the CRCAR was government-controlled and not an intermediary organization as commissions were notionally intended to be.

Alterations in cotton-purchasing arrangements soon followed the creation of the CRCAR. From 1938 importers were required to purchase colonial cotton and were not allowed to purchase foreign cotton until all colonial cotton was placed with a buyer in the metropole. Since these measures effectively eliminated importers' abilities to compete with each other in the buying and selling of cotton and to operate freely in international markets, they formed a *grémio* in 1939. The *grémio* established guidelines and provided protection to importers in a rigorously-controlled environment.

Surprisingly the industry did not similarly respond to government alterations and the organization of importers with a *grémio* of their own; indeed one was not formed until the early 1960s. The textile industry did have its own textiles section within the Associação Industrial Portuense and the Associação Industrial Portuguesa but it does seem strange that one of Portugal's largest and oldest industries was not included in the corporate structure. Nor is there any evidence to suggest that the government encouraged the industry to form a *grémio* or that the various and disparate elements in the industry wanted to form one. What is probable is that the government actually preferred to work with the industry through the old associations and that certain interests in the industry also liked the arrangement. Textile sections in both associations were dominated by owners of larger textile mills who had a long history of lobbying as we described in Chapter 2. It is probable that the government preferred to make personal, informal contacts with these more powerful textile industrialists rather than contend with a potentially powerful *grémio* composed of the entire industry.

The government was able to establish a grip on the rest of the industry by other means—through the importers' *grémio* (GNIAR) and the CRCAR. In 1939 there were only 25 licensed importers, all in the *grémio*, and by 1950 the number had only risen to 38 consonant with the growth of demand.[20] All importers had to belong to the *grémio* and it was probably much easier to control 25 to 38 importers than trying to control an industry composed of

[20] Grémio Nacional dos Importadores de Algodão em Rama, *Relatório e contas*, 1950 (hereinafter GNIAR, *Relatório*).

around 250 factories (455 in 1950). Each importer was allotted a quota of cotton consonant with his imports over the last two years and each had to state with whom he was transacting business and how much cotton he was trading. This enabled the government to follow the cotton purchases of each and every factory in Portugal. Through controlling the distributors of the cotton supply, the government monitored and supervised the textile industry.

World War II greatly accelerated government intervention in the industry. With the start of the war, the purchase of foreign cotton became erratic. Textile industrialists muttered about the cotton supply which seemed to vary not only from year to year but also from month to month. One company noted in its annual statement for 1941:

> Presently we cannot fill during normal business hours the many orders that we have in hand and we also continue to have empty warehouses. Our products are ranked amongst the best, and the value of our shares is up. Nevertheless we ought to remember that the future does not smile on us, principally because of the absolute lack of cotton. At the moment we have on hand an official circular informing us that in the present month of February (1942) there will not be any cotton imports. Many factories have already shut and others will shut if these difficulties cannot be overcome. This means that we need to be very prudent and careful in order to defend ourselves from the bad days that promise to come.[21]

That statement referred to the years of 1941 and early 1942. In 1943 ginned cotton imports were only 18,045 tons compared with 23,040 tons the year before, and the difference was mainly attributable to the decrease in imports of foreign cotton. Imports continued to oscillate during the war period creating uncertainty within the industry and provoking vociferous criticism from manufacturers. In a succinct summary of the past and present dilemmas of the textile industry, the directors of the Alcobaça company declared in their report for 1942:

> Yesterday, it was the problem of placement which created grave and complex difficulties, asphyxiating and absorbing energies and efforts; today it is the problem of acquisition. It is true that transactions are strictly controlled, but in addition the company's administration has had to contend with numerous problems associated with the cotton supply.[22]

[21] Alcobaça, *Relatório*, 1941, 6.
[22] Ibid. 1942, 5.

The erratic foreign cotton supply paved the way for its replacement by colonial cotton. By 1950, 90 per cent of cotton imports were from the colonies. Aided by the war, the government had finally wed colonial cotton to the Portuguese textile industry.

The CRCAR used its powers to intervene in the entire production process—from the supply of colonial cotton to the sale of finished textiles. In 1942, when the government was trying to control prices and production in the domestic market, the CRCAR was put in charge of distributing ginned cotton to factories at fixed prices. In turn, it obliged factories to make cloth according to the quality, type, and quantity stated by the CRCAR. All industries were to register with it, the CRCAR then verified that the cloth had been made.[23] In terms of personnel alone the CRCAR grew from a staff of forty in 1940 to 131 in 1944 to 155 in 1945.[24] Indeed, in a justification of its role, the President of the CRCAR, J. A. Simões de Almeida stressed the importance of corporate superintendence of industries: 'Unity, cooperation, self-direction, hierarchy—all are basic rules of corporate reorganization and all are essential principles of the future life of the cotton industry.'[25] Through the CRCAR and Industrial Regulation, the government decided how those principles would be applied.

INDUSTRIAL REGULATION

Worried that the problems of the past—acute competition and over-production—would reappear alongside increased market share at home and abroad, the government was determined to control the industry. As we noted in Chapter 2, there was an enormous increase in machinery and/or reopenings of factories up to 1935. The government responded to the increase with swift action. In 1936 it advised industrialists that in accordance with Decree-Law 19,354 of 3 January 1931 (the Industrial Regulation law), the Director-General of Industry had to give authorization for any

[23] *Bol. da DGI*, 5: 249 (17 June 1942), Portaria 10, 111.

[24] Lucena, 'Sobre', Part II, 162.

[25] J. A. Simões de Almeida, *A indústria algodoeira e a organização corporativa* (Lisbon: Editorial Império, 1940), 78.

installation of new factories or the replacement of machinery in existing ones.[26]

A year later, Law 1,956 restated and refined the aims and objectives of Industrial Regulation generally and allowed organizations of economic co-ordination to comment on all requests for machinery or expansion. Decree-Law 27,994 that same year gave these organizations sixty days in which to comment on requests and described the procedure for transferring to, or replacing machinery in factories. The passage of the law was a virtual admission that industries had observed neither the legislation nor the procedure and had expanded or reopened without seeking government permission, or by bribing the government to turn a blind eye.[27]

As statistics indicate, the government did manage to strictly control expansion in the textile industry by the mid-1930s. According to the CRCAR, the number of working factories included under Industrial Regulation laws only increased by ten in the five years between 1935 and 1940, bringing the total number of factories to 242. The number of installed spindles in 1935 was 600,426 and in 1940, 667,309. The number of looms increased only slightly by 397 (see Table A13). Of those factories subject to regulations then, growth or reopenings had occurred and machinery had increased but they were very controlled.

Yet, the policies employed by the government to control growth were often misguided, and ultimately harmful to the industry. Contrary to government proclamations, intervention did not eliminate cutthroat competition prior to World War II. In 1937 and 1938 the Alcobaça company mentioned that it had acquired new machinery in order to adapt to a changing market—presumably a market which had expanded due to protectionism or the Spanish Civil War. At the same time, the companies of Alcobaça, Fafe, and Guimarães referred to instability, stagnating sales, and cutthroat competition both within and without the country.[28] As the Guimarães company noted, only increased exports not government intervention ameliorated the poor state of the industry:

> 1937 was one of great instability for commerce and industry. Economic life developed with many difficulties and uncertainties in view of persistent

[26] *Bol. da DGI*, 1: 17 (5 Jan. 1938), 'Estatística Industrial' (9 Jan. 1935); and *Indústria Têxtil*, 1: 1 (Nov. 1936), 20.

[27] *Bol. da DGI*, 1: 1 (15 Sept. 1937), D-L 27,994 (26 Aug. 1937).

[28] See company reports for Alcobaça, Fafe, and Guimarães, 1937–9.

disturbances in the world. . . . The contraction of orders has become very notable and because of this sales have been unprofitable at times. Our neighbour Spain has rescued us from this situation. Since the great commercial fabric centre of Catalonia is inoperable, Portugal has had to supply Spain with basic manufactures. This has resulted in some improvement in this period which has pleased us greatly.[29]

In its reports for 1938 and 1939, this company again mentioned unstable prices and stagnating sales, a predicament corroborated by the trade figures for those years. These examples suggest that in spite of new or larger markets and controlled growth, the problem of cutthroat competition remained unsolved. Even in 1941, the Alcobaça company still felt that competition within Portugal was uncontrolled, but after this date, with sales clearly improving, this company had few complaints. Most companies successfully placed their orders during the war years.

World War II and its aftermath facilitated the increase of government intervention and obscured its effects. Due to the mitigating circumstances of the war, industrialists tolerated, even appreciated government intervention. Simultaneously they failed to recognize the dangers of intervention because they were distracted by the excellent export opportunities brought by the conflagration in Europe. Only in the post-war period did the deleterious consequences of government intervention become obvious.

To meet the foreign and domestic demand occasioned by the war, the industry made numerous requests for additional machinery. Signalling its particular concern with shortages in domestic supply due to the shift of many larger mills to the export market, the government opted for a controlled expansion and modernization of the industry by allowing smaller factories to expand or replace their spindles and looms. Part A of a circular issued by the Director-General of Industry in 1940 announced that the ministry would approve all requests for increased capacity or new machinery from the following categories:

1. The expansion of spinning establishments with less than 3,000 spindles, up to that number;
2. The expansion of mechanical weaving establishments with less than 20 looms, up to that number;

[29] Guimarães, *Relatório*, 1937, 5.

3. The substitution of manual looms by mechanical looms in the following proportions:
   a. 2 manuals by 1 mechanical (when one and the other are narrow, considering narrow to be less than 1,20 meters);
   b. 1 manual by 1 mechanical (when both are large);
   c. 1 manual by 1 mechanical (when the manual is narrow and the mechanical is large);
4. The installation of dyeing and cleaning establishments and expansion or improvement of the same if their present condition adversely affects existing weaving and spinning operations;
5. The installation of preparatory machinery for spinning and weaving as long as their capacity matches the rest of the installations;
6. The replacement of preparatory operating machines as long as the replacement does not lead to requests for more looms and spindles to balance the installations.

Part B stated that all other cases were to be studied separately, these included:

1. Requests for new spinning and weaving factories especially those devoted to the production of fine thread and cloth (of which there was still a shortage);
2. Cottage industries (which had their own laws);
3. The expansion of spinning and weaving units not included in numbers 1 and 2 of Part A;
4. Requests for postponement of time-limits attached to the buying, transference, or legalization of machinery and any other requests not included above.[30]

While the Bill appeared to reveal some loosening of the industrial regulations as applied to the textile industry, only very restricted growth occurred. As can be seen from Part A, only the smallest units covered by Industrial Regulation were to be allowed to expand yet little expansion actually took place in that group. The total quantity of spindles was practically unchanged between 1940 and 1943 (Table A14). With regard to looms, according to Table A15, in the under 30 category, there was an overall increase between 1940 and 1943 of 172 looms to make a total of 1,439 looms in that category. The greatest increase in looms, however, actually

[30] João Dias Rosas, 'Alguns aspectos da situação da indústria algodoeira nacional', Sep. do. *Boletim da Direcção Geral dos Serviços Industriais*, 5: 217–23 (8 Apr. 1953), 5; and *Bol. da DGI*, 3: 140 (15 May 1940), 'Despacho'.

took place in the 31 to 100 category—a size not addressed by the circular—where there was an increase of 896 looms from 1940 to 1943. With regard to the replacement of manual by mechanical looms which was allowed by the legislation, the number of mechanical looms did increase in the period, from 22,286 in 1940 to 23,303 in 1943, but the number of manual looms did not decrease correspondingly as the circular required. In 1943 there were still 732 manual looms—a decrease of only 73 looms since 1940. In addition, the period saw an increase in the use of new automatic looms (215 by 1943) but again this was confined to larger factories using 100 looms or more.[31]

In short, it appears that the ostensible aim of the circular—the expansion and modernization of small weaving establishments—was not realized. João Dias Rosas, later president of the CRCAR, confirmed that neither increases in capacity nor modernization occurred in smaller firms from 1940 to 1943. Rather, increases and modernization could be accounted for by the construction of a new spinning and weaving factory in 1941, to make cloth for export only, which had approximately 10,000 spindles and between 101 and 300 looms. Moreover, he stated that transfers of old machinery from one plant to another rather than the replacement of broken machinery by new machinery were very common and not prevented by the government.[32] The DGI was authorizing larger factories to purchase new machinery and allowing almost all transfers of the old machinery to smaller factories. That large factories were targeted for authorization substantiates Julio Martins's observance of the formation of groups and cliques in the ceramics industry which we cited in Chapter 4; in textiles, the government favoured larger manufacturers. In other words, productive capacity had increased to meet the demands of war—not according to legislative guidelines—but according to bureaucratic preferences and the balance of effective power within the industry. Small manufacturers, however, were allowed to buy the older machinery.

Moreover, company reports confirm that larger factories were allowed to replace machinery or expand in spite of their size. Alcobaça, Fafe, and Porto—all factories with over 10,000 spindles—replaced machinery during the war. The Companhia de

[31] Rosas, 'Alguns aspectos', Map 4.
[32] Ibid. 5.

Fiação e Tecidos do Fafe was authorized to install 2,200 spindles just after the legislation was passed. And in 1942 Alcobaça's Board of Directors stated:

> As is already evident, we need to accumulate all our reserves so that we can adequately re-equip our plants to confront the competition which is bound to arise on all sides, more disturbing than ever, once the first signs of peace appear.[33]

Again in 1943 the company mentioned the need to re-equip in order to face future competition, and in 1945 it declared that it had purchased new machinery and was waiting for transport to resume from Switzerland. A year later the company again discussed re-equipping for the future and by this time other companies had also joined the race to re-equip.

Government action on the question of modernization and expansion took a dramatic turn in 1943 and revealed the extensive grip the regime had on the industry by this time. That year, the government suspended decisions on all requests for the construction of new factories or the expansion of machinery in old factories.[34] Foreseeing the end of the war, the loss of the profitable export market, the meagre possibilities of increasing the internal and colonial markets, and therefore the return of cutthroat competition to the industry, the government refused all requests and took time to consider the situation. From 1943 to 1946, almost no authorizations were given on the 400 requests for replacements or expansion, the granting of which would have resulted in the installation of 523,935 spindles and 12,583 looms.[35] Expansion and modernization of the industry came to a virtual standstill. The number of looms increased only by 269 from 1943 to 1946. Spindles actually decreased by 2,882 (see Table A13).

A study of the industry followed in 1946 and showed that some increases in productive capacity had occurred but without modernization. Regarding growth, the situation in 1946 was the following: there were a total of 254 industrial plants with 669,147 spindles and 24,519 looms (see Table A13). Looking at 1940 to 1946, the total number of plants increased slightly from 242 to 254 and spindles

[33] Alcobaça, *Relatório*, 1942, 15.
[34] *Bol. da DGI*, no. 485 (25 Dec. 1946), 'Despacho normativo' (17 Dec. 1946).
[35] Ibid.

decreased slightly by 4,726. Looms showed an increase of 1,428. Most of this increase was in mechanical and automatic looms. The former rose from 22,286 to 23,624 in 1946; the latter, automatic looms, rose from 82 in 1941 (not available or non-existent in 1940) to 232 in 1946.[36]

With respect to different kinds of plants and their sizes, the greatest expansion had been in weaving only plants, in order to meet the external demand for cloth as opposed to thread, or to replace those mills which had abandoned the domestic market. As Table A15 shows, the greatest increase took place in the category of 31 to 100 looms, which went from 73 mills and 4,245 looms in 1940 to 84 units and 5,152 looms in 1946. However, the statistics did not include nor did the legislation address the existence of cottage industries (estimated at over 1,000 mills with 8,000 looms) which were also competing in the internal market throughout the period. Their existence may explain the persistent references to competition even when the government seemingly had controlled growth. The cottage industry acquired political significance when continuing competition within the industry forced a tightening of the legislation after the war.

Of those regulated, the total number of factories engaged in spinning stayed constant at 89 between 1940 and 1946. The largest factories with over 20,000 spindles recorded no increase. The number of units stayed constant at 6 and their total number of spindles only increased by 126. Some shifting of plants from one category to another occurred as extra machinery was put to use or old machinery retired but very little change occurred overall (Table A14). For factories engaged in spinning, Industrial Regulation succeeded in preventing the establishment of new factories and in controlling the expansion of existing ones.

What became a subject of concern was the nature of the control over expansion for it had overlooked the problems of modernization. Spinning was in a decrepit state. Fifty-three per cent of the spindles were over 30 years old and 13 per cent dated from the nineteenth century; only 5.7 per cent had been bought in the last ten years. In addition, in 1946, 47 per cent of looms in Portugal were mechanical; only 1 per cent of looms were automatic compared

[36] Rosas, 'Alguns aspectos', Map 1.

with 95 per cent in the United States.[37] Essentially the industry had expanded not through the construction of new modern units but by cramming extra equipment into old units; or manufacturers had replaced machinery by buying old equipment from other factories rather than purchasing new equipment. The war obstructed the purchase of new machinery but given that the 1940 legislation had sought to encourage a limited modernization, the high percentage of old spindles and manual and mechanical looms was alarming.

With regard to size, out of a total of 230 factories devoted to weaving only or weaving and spinning, 175 had less than 100 looms and 91 had less than 30 looms. Only 21 factories had over 300 looms, with 49 per cent of the total looms in the industry (Table A15). In England, only 18 per cent of factories had less than 400 looms each whereas 91 per cent of factories in Portugal had less than 300 looms installed.[38]

Spinning establishments were small as well, in 1946 there were 89 factories with spinning facilities. Of those 89, 48 had less than 5,000 spindles; 20 between 5,001 and 10,000; and 15 between 10,001 and 20,000. Only 6 had between 20,001 and 40,000 spindles yet these 6 had 27 per cent of the total number of 669,147 spindles. There were only 21 mills with over 10,000 spindles (including the 6) yet they contained 58 per cent of all spindles (Table A14). By way of comparison, in England only 1 per cent of the total number of spindles were installed in factories with less than 20,000 spindles compared with 42 per cent of spindles installed in factories with less than 10,000 spindles in Portugal.[39]

The picture was bleak. The age and structure of the industry suggested severe problems, not only because the machinery was extremely old, but also because the textile industry in Portugal had a very low concentration of spindles per factory (less than 10,000) and looms per mill (less than 300).

Moreover, the ratio of spindles to looms was low. The national average was calculated at 23.7 spindles per loom in the independent spinning and weaving mills and 28.7 spindles per loom in combined spinning and weaving mills, whereas in England the average was 96 spindles per loom. Even Spain had a slightly higher average

[37] *Bol. da DGI*, no. 485 (25 Dec. 1946).
[38] Ibid.
[39] Ibid.

of 31.5 spindles per loom.[40] This is not even counting the estimated 8,000 looms in cottage mills which had virtually been ignored by the government.

The end of the war then, found the textile industry in a paradoxical situation. Industrial Regulation and the war had allowed existing factories to take advantage of new opportunities and to increase sales. Yet, these two occurrences had prevented the modernization and expansion which were required to reap further benefits from such opportunities. Looms which were old in 1939 were seven years older in 1946; modernization plans of the 1930s were still on the drawing board in the 1940s. On 17 December 1946 the Director-General of Industry, Mr Fausto Carreiro remarked:

> Beyond the problem of quantity, it seems fitting, before anything else, to confirm if production corresponds in quality and in price to technical possibilities and to the legitimate requirements of the consumer.[41]

Clearly production did not. Industrial Regulation had controlled growth but had not improved the industry. Little wonder then that most industrialists viewed the post-war period with trepidation.

[40] Ibid. and see Rosas, 'Alguns aspectos', Map 1.
[41] *Bol. da DGI*, no. 485 (25 Dec. 1946).

6

The Intensification of the Cotton Campaign

Salazar's efforts to assert greater metropolitan control over the colonies while making them pay for themselves appear to have been at least partially successful by the mid- to late 1930s. In the two largest African colonies, Angola and Mozambique, the government balanced budgets by ruthlessly cutting services and encouraged increased trade with Portugal. Trade between the metropole and the colonies served the dual purpose of increasing Portugal's share of the import market and satisfying the regime's fiscal requirements for a positive balance of payments, a healthy foreign exchange, and a strong escudo. Portugal's share of imports into the colonies went from 7 per cent in 1926 to 10 per cent just before the war, peaking at 23 per cent in 1942, and stabilizing at around 14 per cent thereafter.[1]

To this end, centralized direction from the metropole was increased. Whereas before, settler opinion had been consulted in matters pertaining to the colonies, the Colonial Act, the Organic Charter, and Overseas Administrative Reform Act ensured that the metropole would now have the final word on all colonial policies. Along with greater administrative control came a clear preference for development of the colonies so as to directly contribute to the enrichment of the mother country. In 1936 central government passed a general decree-law prohibiting the creation in the colonies of any industry which directly competed with a metropolitan industry; only those which reduced expensive foreign imports or increased the value of exports to foreign countries were allowed.[2] A year later, the passage of Decree-Law 27,552 extended the corporate system to the colonies. The formation of *sindicatos*, *juntas*, and commissions strengthened metropolitan control over

[1] Clarence-Smith, *The third Portuguese empire*, 147.
[2] Ibid. 164.

colonial activities; and the assignment of specific commodities to be produced by the peasants of each colony further increased metropolitan direction of their economies. Cotton was assigned to Mozambique, beans and castor oil to Angola, and ground-nuts to Guinea.[3] Finally, corporate bodies were established in 1939 to control imports and exports and to give priority to Portuguese products. The colonies were to serve the metropole. The Minister of the Colonies, Vieira Francisco Machado summed up the general attitude when he said: 'Portugal for the Portuguese (and) Mozambique for Portugal'.[4]

THE COTTON SYSTEM

Having consolidated its power and committed itself to economic autarky by the mid-1930s, the government returned to the problem of the cotton supply. As we have seen, incentives to cotton concessionary companies in 1932 had certainly improved production but in 1936 the metropolitan industry still imported 71 per cent of its cotton from abroad. Internationally prices were low due to high yields in most of the major cotton-producing countries—Brazil, Egypt, China, India, and Russia—and due to United States government subsidies to producers which lowered the export price of American cotton. The industry was also worried about the quality of colonial cotton and seemed reluctant to switch to colonial suppliers for the reasons cited in Chapter 3.

As with the establishment of the CRCAR in the metropole, greater organization was achieved through the creation in 1938 of the Junta de Exportação do Algodão Colonial (JEAC). Like its metropolitan counterpart, the CRCAR, the purpose of the JEAC was technical and supervisory. Its role was to increase production and improve the quality of cotton. This was undertaken by distributing seed to Africans, supervising its cultivation (alongside colonial administrators and concessionary company managers), and organizing the state markets where companies bought the cotton crop from producers. The JEAC kept a record of all cotton growers,

[3] Vail and White, *Capitalism and Colonialism*, 274.

[4] Dr Vieira Francisco Machado quoted by Carrie Rothkugel, 'The Good Neighbour', *Libertas*, 3: 8 (July 1943), 29.

exhorted growers to intensify cultivation, and established guidelines for the improvement of cotton quality and output. The JEAC also set up experimental stations (16 in Mozambique, 13 in Angola) to conduct seed studies and to test cotton fibres. It instituted and updated a system of classification for qualities of colonial cotton. Most of these activities had been and were being tried by Portugal's imperial neighbours but few lasted as long or had such extensive state involvement as those in the Portuguese possessions.

The technical resources of the JEAC were improved in 1943 with the creation of the Centro de Investigação Científica Algodoeira (CICA). Created by the Colonial Minister and the CRCAR and financed with 10,000 contos (proceeds from a tax on foreign cotton), the Centre's job was to increase yields, discover suitable areas for cotton growing, and improve quality by experimenting with different cotton varieties.[5]

In its efforts to improve the quantity and quality of production, the JEAC worked alongside and with the concessionary companies. Initially, it consolidated the rather haphazard arrangement of cotton zones in both colonies. Angola had 18 small concessionary zones by 1946, many of which were located in the north-eastern part of the country in the area around Luanda. The most important region was Malange but Ambriz, Porto Amboim, and Catete were cotton-growing areas also. Cotonang, whose origins we have discussed previously, was the largest of the cotton companies in Angola. By the 1940s Cotonang had 5 zones and 13 ginning factories, supervised 35,000 Africans, and produced 8,000 to 10,000 tons of unseeded cotton per year—65 to 75 per cent of Angola's total cotton production.[6] Other prominent cotton concessions included Fomento Geral de Angola, Algodoeira Colonial Agrícola and Lagos e Irmão (see map 2).

Mozambique had twelve large concessionary zones most of which were located in the central and northern parts of the country. There was also a small area in the south near Lourenço Marques (today known as Maputo). Amongst the largest concessionary companies were the Companhia dos Algodões de Moçambique (CAM) and the Companhia Nacional Algodoeira. CAM was the

[5] E. K. de Queiroz Ribeiro, *O algodão: Da colheita à industrialização* (Oporto: Imprensa Portuguesa, 1946), 160.

[6] Henrique Galvão and Carlos Selvagem, *Império ultramarino português*, iii. *Angola* (Lisbon: Empresa Nacional de Publicidade, 1952), 360–1.

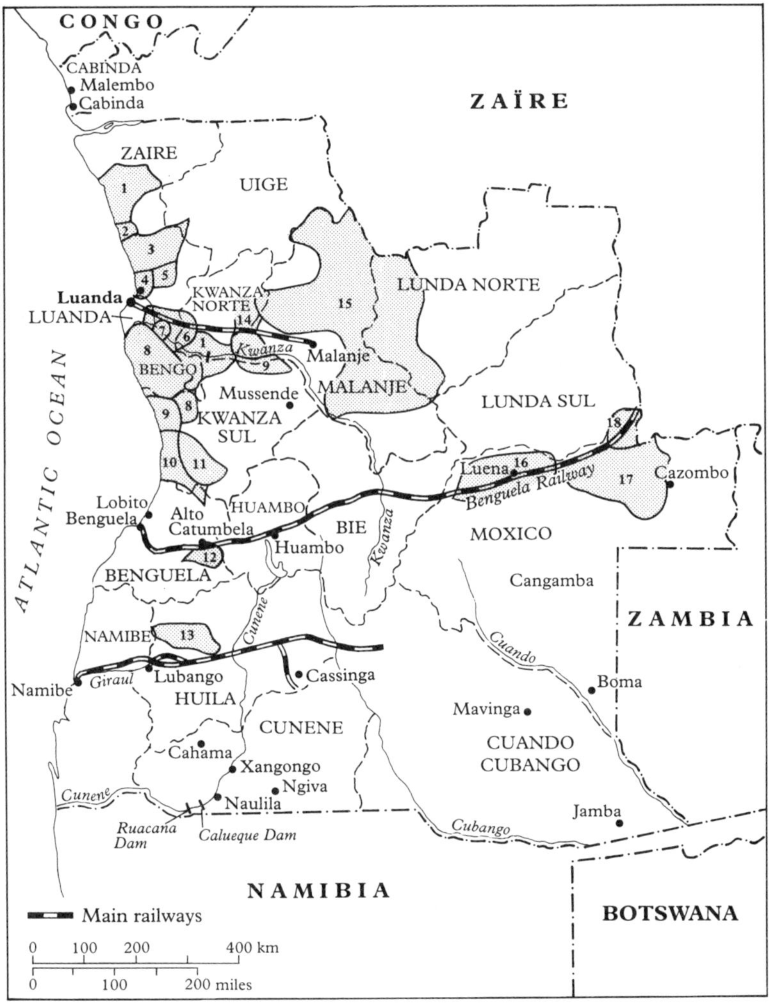

1 Soc. Comercial do Ambriz, Ltd.[a]
2 Soc. Agricola do Cassequel
3 C.[a] do Ambriz
4 Soc. Agrícola e Industrial 'La Luinha'
5 Empresa dos Tabacos de Angola
6 Lagos & Irmão
7 Soc. Comércio e Construções
8 Rocha Coelho
9 Algodoeira Colonial Agrícola
10 Marques, Seixas & C.[a], Ltd.[a]
11 Algodoeira Colonial
12 José Simões Coelho
13 Pereira Simões & C.[a]
14 Soc. Ag. e Com. dos Alg. e Tabacos Lucalo, Ltd.[a]
15 C.[a] Geral dos Algodões de Angola
16 C.[a] Cultora do Algodão do Moxico
17 José Caetano Galvão
18 C.[a] Algodoeira do Dilolo, Ltd.[a]

MAP 2. Cotton Concessionary Companies in Angola, 1946. (Adapted from Joseph Hanlon, *Beggar Your Neighbours*; overlay adapted from JEAC (1946).)

largest of the concessions in Mozambique with 22 zones, 12 ginning factories, and 45 per cent of the total production in Mozambique. Two princes of the Luxemburg royal family, Louis de Burbom-Parma and Felix de Luxemburg, originally claimed the concession in 1920. Previously constituted as GRANDUCOL or the Société Cotonière Luso-Luxembourgeoise, the company really only began supervising cotton production in 1930 with a starting capital of 100,000 contos subscribed in Portugal and Luxemburg.[7] Company directors engaged in notorious labour practices during the 1930s to get Africans to produce, but it was not this that lost the concession. By 1943 the company was badly in debt, Luxemburg had been taken over by Germany, and the nationalist message of the Portuguese government was at its height. Buoyed up by the Law of Nationalization of Capital, Portuguese shareholders lobbied successfully to get the company transferred. It was officially constituted in 1943 when Lagos e Irmão, a cotton concessionary company originating in Angola, bought it from Banco Nacional Ultramarino after paying off its 90,000 contos debt.[8]

Another large concession, the Companhia Nacional Algodoeira consisted of 9 cotton zones formerly belonging to the Mozambique Company which were formed into a separate company in 1943 when the charter of the Mozambique Company expired. Its initial capital was 60,000 contos and it comprised 21,000 hectares.[9] Other companies such as Sena Sugar Estates, the Zambezia Company, and the Companhia Colonial do Buzi, also had small concessions devoted to cotton which were in addition to their other agricultural activities (see Map 3). Neither in Mozambique nor in Angola is there any evidence that metropolitan textile companies invested in these cotton concessionary companies—indicative of the scepticism with which they viewed colonial cotton production.

Along with the JEAC, the concessionary companies were in charge of distributing seeds to Africans and supervising the cultivation of cotton. The companies had their own overseers and *cypaes*

[7] Henrique Galvão and Carlos Selvagem, *Império ultramarino português*, iv. *Moçambique, India, Macău, Timor* (Lisbon: Empresa Nacional de Publicidade, 1953), 192–3.

[8] Carlos Fortuna, 'Threading Through: Cotton Production, Colonial Mozambique and Semi-Peripheral Portugal in the World Economy', Ph.D. thesis (SUNY-Binghamton, 1989), 223.

[9] Galvão and Selvagem, *Império ultramarino*, iv. 194.

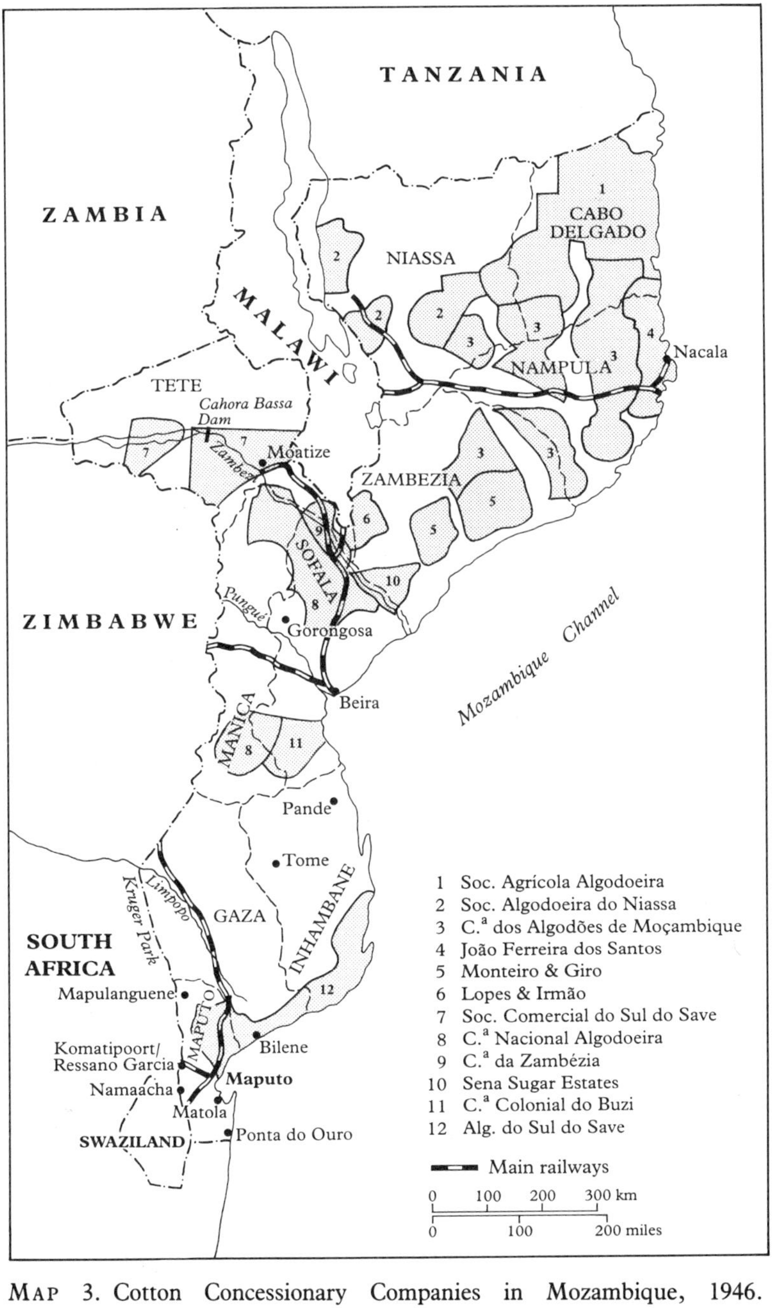

MAP 3. Cotton Concessionary Companies in Mozambique, 1946. (Adapted from Joseph Hanlon, *Beggar Your Neighbours*; overlay adapted from JEAC (1946).)

(African soldiers in government service) for this purpose, but also they had their own experimental stations and agronomists. The large concessionary company in Angola, Cotonang, even had its own variety of cotton, 'Triunfo', which it tested and perfected over the years. After the cotton harvest, companies were responsible for ginning unseeded cotton bought from African producers and exporting the ginned cotton to Portugal.

In conjunction with the JEAC and the concessionary companies, the colonial administration was expected to aid the scheme. Together, the government and the JEAC fixed prices for the cotton. The markets where producers sold their cotton to the concessionaires were government run and supervised by local administrative authorities and the JEAC. Moreover, since private companies were forbidden legally to compel Africans to grow cotton under the 1928 Labour Code, administrators of *concelhos* (councils), *chefes de postos administrativos* (heads of administrative posts), and the JEAC were responsible for encouraging Africans to grow cotton and ensuring that they harvested their cotton crop.[10] But officials engaged in much more than simple encouragement. Owing to the loophole in the 1928 Code, they could compel Africans to work, and after 1936 forced labour became the rule rather than the exception. Under threat of punishment, Decree 27, 552, the same decree that extended the corporate system to the colonies, required Africans to grow crops which would contribute to their nutritional improvement, or could be sold for cash, or could be exported.[11] Cotton of course fell in the latter two categories.

To obtain sufficient labour, administrators and the JEAC relied on African chiefs to exhort their subjects to produce cotton, and the JEAC, administrators, and chiefs began co-operating with companies to force Africans to grow cotton.[12] In 1942, with the increased demand for raw materials generated by the war, a circular then legally sanctioned what was already widespread practice: the use of forced labour and the participation by companies in recruitment. The circular declared that

[10] Vail and White, *Capitalism and Colonialism*, 276.

[11] Silva Cunha, *O trabalho indigena* (Lisbon: Agencia Geral das Colónias, 1949), 174.

[12] For the role of chiefs in the cotton scheme, see Isaacman, 'Chiefs, Rural Differentiation', 15–56.

all Africans between the ages of 18 and 55 were to be obliged to prove that they lived on the proceeds of their labour, and that they had worked for at least six months of the year as labourers in the service of the state, private persons or companies.[13]

In addition to directly contravening the 1928 Labour Code, it produced a scramble among companies for labour, and sowed confusion among administrators as to their loyalties. Particularly in Mozambique, where cotton had been assigned as the principal commodity to be grown by peasants and where, because of this, the JEAC was most active, recruitment and forced labour produced numerous conflicts.[14] We shall return to these conflicts briefly.

Due to supply shortages as a result of the war, interest by the metropolitan textile industry and corporate organizations was greater by the early 1940s so that they too were involved in the cotton process. In addition to its metropolitan duties, the CRCAR worked closely with the JEAC not only in the setting of prices and the classification of cotton but also bought cotton when, as in 1941, there was not sufficient demand by importers. Government cotton organizations, the cotton concessionary companies, and the colonial administration thus joined forces to increase and improve, regulate and supervise colonial cotton production. The process was difficult and detailed and merits description in order to understand the aims of such official involvement.

PROCESS OF COTTON CULTIVATION

Cotton is an annual crop and new seeds must be sown each year. This has the advantage to producers of allowing them easily to switch to food crops or more profitable crops for the year if cotton prices are low. This may explain the initial leaps and drops in production figures. It also explains why supervision was so necessary.

Once embarked upon, cultivation is a tedious process. Careful attention must be paid to the amount of soil covering each seed, the distance between plants, and the time seeding takes place. Picking was crucial, of course. The Director of Agriculture for the

[13] Vail and White, *Capitalism and Colonialism*, 280.
[14] Ibid. 276.

Mozambique Company exhorted concession holders in 1939 to encourage Africans to harvest 'when the bracts of the bolls appear dry and brown, and the fibre hangs freely from the loculus'.[15] Otherwise, if the crop was harvested too early the fibres would be too short and too dense, and if too late, dry, brittle, and damaged.

Other fears concerned levels of humidity, not only where the cotton was grown but also where it was stored after ginning. If the cotton was too wet, it could collect dirt or rot. These concerns also extended to the sacks used to collect the unseeded cotton and the baling method used for ginned cotton. Apart from these hazards, those involved in cotton production had to worry about the ever present boll-weevil, the destroyer of not just fields but an entire nation's crop, as in the USA in the 1920s.

As stated in the 1926 Decree-Law 11,994 concessionary companies operating in specified zones in Angola and Mozambique organized the scheme. The zones were supposed to be in areas declared suitable for cotton growing but subsequent events have shown that this was not always the case. Zone sizes and numbers varied depending on the capital of the concessionary company, its interest in cotton, and the changes in legislation. Some companies were solely devoted to cotton (more typical in Mozambique, since cotton was the designated crop for that country) whereas others grew cotton in addition to sisal, copra, or sugar (more descriptive of the Angolan situation). Mozambique had fewer concessionary companies than Angola, but almost all had substantial holdings. For example, CAM had 23 concessions in three different provinces of Mozambique solely devoted to cotton production whereas the Sociedade Agrícola do Cassequel only had a single small concession in Angola for cotton cultivation in addition to its extensive sugar concessions. The 1926 legislation had put the maximum size of zones at 50 kilometres each but Decree-Law 20,881 of 1932 extended the maximum to 120 kilometres for the size of each zone.[16]

The concessionary company responsible for each zone sent overseers to mark out a *machamba* or cotton-growing area for each African according to criteria drawn up by the governors of each province. It has been estimated that it takes approximately 140 to

[15] Ribeiro, *O algodão em Moçambique*, 19.
[16] Ibid. 63; and Art. 37, D-L 11,994 of 1926.

150 days to tend by hand 1 hectare of cotton; and 37 per cent of that time is spent collecting the crop by hand.[17] If food crops are grown as well, then cotton ought not to occupy more than half a hectare. In 1941 the average plot of cotton cultivated was a third of a hectare in Mozambique and nearly a hectare in Angola. By 1945 the average size in Mozambique had increased to two-fifths of a hectare but sizes of individual plots tended to vary widely.[18]

Once the cotton was picked, it was then taken to markets to be sold to concessionary companies. All buyers who entered the market had to be registered with the JEAC. These markets were strictly controlled by local administrative authorities and the JEAC in order to prevent competition from non-registered buyers. Moreover, as little transport was available and most Africans had to walk, the 1926 decree stipulated that markets not be located more than two days' walk from the cotton zones. The Governor of each colony decided when and where cotton markets would be located and how many days they would stay open.[19] The 1926 legislation was quite specific about daily closing times, stating that markets should close 'an hour before sunset'.[20]

The names of all producers were written in market registers including the quantity and classification of the producer's cotton as well as the price paid for it. Old people and children were often sent to market to sell the cotton but were counted as one producer, an indication that many more people were involved in cotton growing than the administration admitted.[21] Producers were paid at the end of the day or at the end of each market at a central cashier's office which handled all the transactions of the cotton concessionary companies or the JEAC, depending on who bought the cotton. If the crop were larger or better than anticipated and money ran short, companies or the JEAC contributed more and if any money remained, it belonged to the administration and was 'put to the assistance of Africans'.[22]

The cotton which was brought to market also had to be classified

[17] Nelson Saraiva Bravo, *A cultura algodoeira na economia do norte de Moçambique* (Lisbon: Junta de Investigações do Ultramar, 1963), 202; Ribeiro, *Da colheita*, 95.

[18] Bravo, *A cultura algodoeira*, 130.

[19] *Diário do Governo*, D-L 11,994, Art. 20.

[20] Ibid. Art. 24.

[21] Bravo, *A cultura algodoeira*, 134–5.

[22] Ibid. 133.

according to quality. Regardless of the type of seed used, the colour and length of cotton varies from country to country and even region to region depending on climactic conditions, the quality of the soil, the method of cultivation, and the manner of harvesting. Since some standard of comparison must be used when ginned cotton is traded on the international market, a universal classification system for cotton was adopted in 1923.[23] According to this system which was periodically revised, standard American middling—the grade of ginned cotton most often bought by textile manufacturers—was adopted as the universal standard of comparison. Other countries could name and classify their own cotton as they wished but it was always graded in relation to standard American middling. This allowed purchasers all over the world to buy similar grades of cotton at standard prices. Even the Portuguese, who did not conform to international prices, used standard American middling as a guide when setting prices and devising their own classification system.

When cotton production first began in the Portuguese colonies, the quality of the cotton was extremely poor. Fibres were irregular and dirty, the colour yellow. Initially, the JEAC classified both colonial unginned and ginned cotton into three grades—1st, 2nd, and 3rd quality—depending on such factors as cleanliness of the cotton, the length of the fibres, and the colour.[24] But with steady improvement in methods of cultivation, supervision, ginning, baling, and storage, the percentage of 1st quality improved and it became possible to make finer distinctions between different qualities of cotton. In 1945 the JEAC altered the classification of ginned cotton to a 6-grade classification system. Unginned cotton retained a 3-grade classification system.

When Africans brought the unginned cotton to market, it was classified and the Africans paid according to whether it was 1st or 2nd quality (3rd quality was virtually unusable). This was a particularly tricky, but extremely important stage, because if cotton was classified incorrectly before ginning, then it became difficult to predict the overall quality for the year, and hence, whether there

[23] Ribeiro, *Da colheita*, 132.

[24] Both Vail and White ('"Tawani, Machambero!" Forced Cotton and Rice Growing on the Zambezi', *Journal of African History*, 19: 2 (1978), 250) and Bravo (see example below) cite evidence of companies doing the classifying of unginned cotton even though officially the JEAC was responsible.

TABLE 6.1. *Percentages of Unginned and Ginned Cotton in a Random Sample, 1942–1943*

Quality	Unginned	Ginned
1st	30%	40%
2nd	50%	45%
3rd	20%	15%

Source: Ribeiro, *Da colheita*, 108.

would be enough of the right quality cotton for the industry. Complaints by the industry suggest that this happened and that cotton of the right quality was therefore not ordered from elsewhere in time to satisfy the industry. Moreover, if the cotton was graded 1st quality unginned but 2nd quality when ginned, then the cotton concessionary companies lost money on the transaction. This worked the other way too. One writer noted that cotton produced by Africans in Angola during the early 1940s was consistently classified as 2nd quality when it should have been classified as 1st. A random sample conducted in 1942–3 shows the relation of unginned to ginned cotton in Angola (see Table 6.1).

The figures in Table 6.1 suggest that Africans were paid even less than they were entitled to, due to the incompetence or dishonesty of government classifiers. Official statistics on the quality of unginned and ginned cotton actually confirm that a discrepancy existed between the classification of unginned and ginned cotton in most years up to 1945. The figures in Table 6.1 also challenge official percentages of 1st and 2nd quality ginned cotton in Angola which were given as 80 per cent for 1st quality and 12 per cent for 2nd in 1942; and 74 per cent and 12 per cent for 1943 (Table A12). In addition, a cotton expert, N. S. Bravo observed that as late as 1956, 'the personnel of some businesses show evident partiality towards their employers by buying 1st quality cotton at 2nd quality prices'.[25] Official statistics, the random sample, and Bravo's testimony suggest that the persistent complaints by importers and manufacturers about the poor and variable quality of imported

[25] Bravo, *A cultura algodoeira*, 181.

colonial cotton were probably justified and that Africans were cheated by concessionary companies.

The colonial Governor in consultation with the JEAC fixed the prices to be paid to African producers at the market. Once payment took place, the cotton was then taken to be ginned and baled in factories owned by the companies. The law required all concessionary companies to have their own ginning factories in order to prevent resale on parallel markets. In Angola, there were 27 ginning factories with 7 auxiliary offices and in Mozambique, 26 with 21 offices in 1943.[26] Most of the gins were old, having been bought second-hand. Their inefficiency tended to lower the ratio of ginned to unginned cotton (which ideally should be 3 to 1) and also affected the quality of cotton obtained. In principle, this also depressed profit margins for the companies, but one Mozambican company reported that the government paid a subsidy if the percentage of cotton left after ginning did not equal 32 per cent of the original amount. This company consistently failed to attain the 32 per cent ratio, an occurrence which was not at all uncommon well into the 1950s.[27] It may be that it was in a company's interest to report a lower percentage and collect the government subsidy. Yet this same company equally complained that often the government failed to pay the subsidy.

Baling and pressing were also often done on second-hand machines with the result that bales were of various sizes. This tended to prejudice selling in Portugal and to alter the statistics. Big bales tended to be too heavy and to affect the cotton quality. With small bales, which used more baling wire and twine, costs rose.[28] Small details such as irregular baling and ginning all tended to affect the quality and price of colonial cotton.

Despite the inherent difficulty of the exercise, the JEAC attempted to equate the types of ginned Mozambican and Angolan cotton with the American standard. In 1946 Mozambique 1 and Angola 2 were equated with American Good Middling, Angola 4 and Mozambique 4 with Standard American Middling. Some of Angola 1 was of exceptionally high quality and production was limited to a few regions. The Triunfo variety produced by Cotonang

[26] Ribeiro, *Da colheita*, 271.

[27] Companhia Nacional Algodoeira, *Relatório*, 1951, 1953, 1956, and 1961; and Fortuna, 'Threading Through', 248).

[28] Ribeiro, *Da colheita*, 125.

attained this classification and was used by the textile industry for the production of very fine cloth only. Notwithstanding the success of Triunfo, however, constant complaints by domestic textile manufacturers suggest that the classification system, in addition to classifying most of the cotton as Type 1, also overrated colonial cotton in comparison with its American equivalent.[29] The majority of cotton tended to be classified as Type 1 and consequently Type 1 was the variety most often purchased, but it frequently did not conform to the international standard.

PRICES OF COLONIAL COTTON

Once classified, cotton concessionary companies then sold the cotton to metropolitan importers. There has been confusion not only as to what prices were paid for cotton during the colonial period but also to whom they were paid. Several scholars of Portuguese colonialism have erroneously concluded that cotton prices were always set below world levels to benefit the Portuguese textile industry.[30] Other interests—those of the government, the concessionary companies and the African producers—were involved, however, and in two periods, the 1930s and the late 1950s, the pressures exerted by these interests actually led to prices which were above world levels. Yet even this needs to be stated with qualifications: for subsidies were often granted but not paid, and premiums were suddenly dropped.

As can be seen from Table A7, the set price of 8$00 escudos in 1932 was 67 per cent higher than the international price at the time. As we noted in Chapter 3, metropolitan importers bought cotton from concessionary companies at the world price and the government then reimbursed the companies the difference between the set price and the world price. The objective of this scheme seems to have been mainly to offer incentives to cotton conces-

[29] High-quality cotton production is limited to a very few areas in the world; therefore it is unlikely that Portuguese colonial cotton would have surpassed American middling in quality. Most likely, Type 1 was more or less equal to American middling, hence the comparison of American middling with Portuguese Type 1 in Table A7.

[30] See Newitt, *Portugal in Africa*; Munslow, *Mozambique*; and Isaacman and Isaacman, *Mozambique: From Colonialism to Revolution*.

sionary companies but as we pointed out, the government did not always pay the subsidies. In 1936 importers paid the world price of 6$38 per kilo as before, but the government dropped the premium paid to concessionary companies. World prices for American middling dropped to 4$40 per kilo in 1937 and the same price was paid for 1st quality colonial cotton from the Portuguese colonies, without the premium to cotton concessionary companies. The following year, the government then reinstated the grant to bring the total paid to concessionary companies to 7$50 per kilo in 1938, even though the world price was only 4$55 per kilo (Table A7).

At this time, the system of reimbursement was placed under control of the CRCAR. Moreover, importers rather than exporters were to be reimbursed. These changes placed responsibility as well as control in the hands of the metropole rather than the colonies, additional evidence of metropolitan centralization in this period. Furthermore, the method of reimbursement tied importers financially to the CRCAR. Importers were obliged to pay a government fixed price for the purchase of colonial cotton and to pay colonial exporters within eight days of purchase. The CRCAR agreed to reimburse the importer the difference between the international price of an equivalent quality of cotton and the price of colonial cotton if the price of colonial cotton was higher on the day of arrival at the ports of Lisbon or Oporto. Reimbursement of importers was achieved by a tax on foreign cotton of 60 centavos per kilo.

The main objective in switching reimbursement to importers appears to have been to maintain stable prices to cotton exporters—a government goal—and so to continue to offer incentives to cotton concessionary companies at a time when the government was particularly keen to tie the metropole and the colonies together.[31] This arrangement, though periodically revised, lasted until 1942. When world prices started climbing due to the war, the government decided to re-fix prices in 1942 to avoid the destabilizing effects of oscillating prices. It was at this time that colonial prices were set below world prices but the differences were not great until the end of the war (see nominal and real prices, Tables A7 and A8).

Whether the export price was above or below international prices of cotton, few benefits accrued to producers given the low outputs

[31] Ribeiro, *Da colheita*, 339.

and low producer prices for the period. The average producer price per kilo for 1st quality dropped after 1939 from 98 centavos to 95 centavos in 1940 and 1941. Due to complaints by African producers and criticisms by local administrators, the colonial government in Mozambique tried to raise the average producer price to 1$25 escudos in 1942, but concessionary companies (who by this time were not receiving prices comparable to international prices and were afraid their profit margins would drop) pressured the administration to decrease them the following year.[32] Even given the price rise to 1$20 achieved in 1943, the income for producers was hardly remunerative (see Table A10, real producer prices). Whereas in Mozambique a hectare of peanuts yielded approximately 700 escudos in 1944, a hectare of cotton brought at most 200 escudos.[33] Most producers farmed less than a hectare, therefore the income per producer from cotton was even lower. Since Africans had few monetary incentives to produce cotton, companies and the administration resorted to force. Constant supervision, threats, and beatings were common tactics in the struggle to supply the metropole.

DISTRIBUTION TO TEXTILE FACTORIES

Once these tasks were accomplished, the cotton was ready for shipment to Portugal. The establishment of the CRCAR and its colonial counterpart, the JEAC, provided the government with the organizational support to tie colonial production and metropolitan purchase more closely together. In 1938 Decree-Law 28,851 altered the method of colonial cotton purchases and their distribution. All concessionary companies were required to declare their production and only that which was declared could be exported. All cotton concessionary companies had to belong to the JEAC and required a JEAC certificate to sell cotton to the metropole. All transactions had to be made with metropolitan importers registered with the CRCAR. If no buyer for the colonial cotton could be found, the cotton was sold to the JEAC, exported to the metropole, and then distributed obligatorily to importers by the CRCAR. Any importers who refused to accept the mandatory cotton distributed by the

32 Vail and White, '*Tawani Machambero!*', 251.
33 Hedges and Rocha, 'A reestruturação', 57.

CRCAR could be suspended from importing for three to ten months.[34]

In 1938 the government suspended all importers' contracts to buy foreign cotton until the output of colonial cotton was known. Importers were forbidden to purchase cotton unless they had a buyer. All contracts to buy and sell cotton had to be approved by the CRCAR and the commission had to give approval of the final destination of the cotton. All cotton, not only that handled by the CRCAR, but even that bought by importers was distributed according to quotas based on the amount importers had bought over the last two years. All colonial cotton had to be purchased first before importers were allowed to buy abroad. This requirement to purchase colonial cotton in effect prevented importers from exercising futures options on foreign exchanges because they had to wait for the outcome of the colonial cotton campaign to know if their quota had been filled by colonial cotton.[35] In sharp contrast to the relatively free environment in which they had worked before, the importers were now caught in a web of government regulations and approval. During the period when Portuguese colonial cotton was higher than the international price, one importer protested:

> In the first place, no importer and much less any American exporter or cotton producers of other countries have ever compelled the industry to import an inferior product at higher prices (than the international one for the same quality).[36]

The purchasing arrangements for colonial cotton created considerable resentment amongst importers. On the international markets, the existence of spot and futures markets for the purchase of foreign cotton allowed an importer to buy as much cotton of the right quality as he wanted, when he wanted it. The trouble with colonial cotton was that first, the importer had to pay for it on arrival and not when he sold it to an industrialist, and secondly, the industrialist had to take it when it was available rather than when he needed it. This placed financial hardship upon both importers and industrialists and interfered with the production of textiles.

[34] *Bol. da DGI*, 1: 45 (20 July 1938), D-L 28,851.

[35] *Bol. da DGI*, 1: 38 (25 May 1938), D-L 28,698; and *Bol. da DGI*, 1: 45 (20 July 1938), D-L 28,851.

[36] Fred Kjolner, 'Política de algodão', repr. in *Indústria Têxtil*, 3: 22 (Mar.–June 1939), 86.

No less resentful were textile manufacturers. In response to the controlled environment in which they now worked, importers had formed a *grémio* which attempted to: avoid the build-up of stocks, set import quotas for each importer, establish uniform prices, secure a 5 per cent profit for importers, and avoid competition. In seeking to limit the number of cotton importers and to set prices, the *grémio* offered obvious advantages to those already engaged in the business but potential disadvantages to industrialists accustomed to shopping around for the best quality, most cheaply priced cotton. When the idea of an importers' *grémio* was first suggested in 1937, one manufacturer recognized that:

> If the *grémio* is created, free competition between importers, so beneficial for industrialists and even for importers (taking into account average purchases), will disappear. It is a basic rule of trade that buying large quantities means better prices. Precisely because of this, it is unfair that a manufacturer who buys 200 bales should pay the same price as one who buys only 25![37]

Statistics on cotton imports alone show that importers and manufacturers were not allowed to purchase cotton freely until nearly the end of the authoritarian regime. Almost as soon as the CRCAR and the JEAC were established, colonial cotton imports shot up. By 1942, 81 per cent of the cotton was coming from the colonies and colonial cotton comprised the majority of cotton imports for most of the next thirty years (see Table A4). That same year, 1942, the policy of premiums was dropped for cotton concessionary companies and a fixed price instituted. According to the government, this was because international prices were oscillating wildly and price stability was more important than a low or erratic price. However, since the fixed price was now below international prices, it seems, rather, that the government had achieved substantial control over cotton production and distribution by this date.

THE DRAWBACKS OF THE FORCED COTTON REGIME

The dramatic increases in cotton after 1936 hid the appalling conditions in which many Africans produced and the resistance and

[37] X, 'Algumas considerações oportunas acerca da fundação do "Grémio dos Importadores de Algodão"', *Indústria Têxtil*, 1: 3 (Jan. 1937), 15.

conflict which dogged the scheme throughout the 1930s and 1940s. Zealous attempts by concessionary companies to grow cotton at the expense of food crops proved disastrous on a number of occasions, encouraged African resistance, and provoked labour shortages. Food shortages in Mozambique were common during the 1940s.[38] In the cotton concession of the Companhia Colonial do Buzi a famine occurred in 1940 and the company had to distribute food to Africans.[39] In the Zambezia region during the 1940s, African women, the majority of the cotton growers in this area, had to buy as much as 6,000 tons of extra maize from the concessionary company in charge of the area, Sena Sugar Estates.[40] More often than not, the dangerous balance of food versus cash crop appears to have been tipped in favour of cotton.

The demands of cotton growing, pressure, and abuse by colonial administrators, JEAC officials, and company overseers were not met passively but actively resisted by Africans who devised numerous ways of avoiding cotton work. At first, when administrative roles were not yet defined, resistance took the form of simply ignoring the demands. As this became more difficult, Africans in both Mozambique and Angola had a number of options they could pursue instead of cotton cultivation. They could work on other plantations, such as sugar in Mozambique or coffee in Angola. They could emigrate permanently to other countries or to the cities; or they could migrate to work in the mines of South Africa or Rhodesia.[41] Some simply refused to grow the cotton because (it was explained), 'they cannot eat it'.[42]

Resistance could also be more violent. Africans resisted by burning seeds, planting late, or not destroying the previous year's crop properly. Because cotton was an annual crop all sorts of things could happen before the seeds hit the ground.[43] Where overt resistance was limited, psychological resistance persisted and is

[38] Portugal, Agência Geral das Colónias, *Boletim Geral das Colónias*, 22: 250 (Apr. 1946), 165.

[39] Fernando dos Santos Graça, *Companhia Colonial do Buzi* (Lourenço Marques: Imprensa Nacional de Moçambique, 1942), 138–9.

[40] Vail and White, 'Tawani Machambero!', 253.

[41] Judith Head, 'Sena Sugar Estates and Migrant Labour', in *Mozambican Studies*, no. 1-1 (1980), 61; Pitcher, 'The Agrarian Question', ch. 4; and Douglas L. Wheeler and René Pelissier, *Angola* (London: Pall Mall Press, 1971), 138–9.

[42] Ribeiro, *Da colheita*, 159.

[43] Isaacman *et al.*, 'Cotton is the Mother of Poverty', 595–604 for more details.

evidenced by the numerous songs and drawings criticizing administrators and overseers. In northern Mozambique, Makua and Makonde artists ridiculed state officials and Europeans in carvings which distorted their features and made them sub-human. Women in the Sena area also had a number of songs relating to cotton and its abuses.[44]

If cotton concessionary companies conflicted with Africans they also conflicted with other companies and with the colonial administration over the use of labour. To be sure, cotton had a ready market in Portugal, but other products such as sugar, copra, coffee, and corn also brought in valuable foreign exchange. Some companies, anticipating the conflicting claims on labour or possible high profits to be gained from cotton, had taken out cotton concessions when they were offered, even though these companies' major interests might be in sugar or copra. For example, the Sociedade Agrícola do Cassequel in Angola and the Companhia Colonial do Buzi in Mozambique were mainly concerned with sugar production and not cotton. Taking out the concession was a way of reserving male labour for the more profitable sugar plantations. The Companhia Colonial do Buzi admitted as much in a 1942 monograph when it stated that only female Africans had been advised to grow cotton 'except in cases where male Africans were determined to be cotton farmers'.[45] In 1941 the Buzi company calculated that approximately 18,216 women were available for cotton work. In 1943 the total unseeded cotton yield for the area of the Buzi cotton concession was 1,787,471 kilos. Taking 100 kilos per producer as the average output at that time that would make approximately 17,875 producers.[46] A strike several years later confirms that many of these producers were probably women. Seven thousand women refused to accept seeds handed out by a Buzi manager complaining that they had neither the time to plant nor the hands to harvest cotton. The company backed down and exempted pregnant women as well as women with more than four children from the compulsory cultivation of cotton.[47]

[44] Vail and White, 'Tawani Machambero!', 252; and *Capitalism and Colonialism*, 316–17.

[45] Graça, *Buzi*, 39.

[46] Isaacman *et al.*, 'Cotton is the Mother of Poverty', 588; and Bravo, *A cultura algodoeira*, 131.

[47] Isaacman *et al.*, 'Cotton is the Mother of Poverty', 603. It would be interesting and not at all surprising if birth rates in that area jumped dramatically from 1948.

The use of female labour for cotton growing was also practised by Sena Sugar Estates which in 1942 embarked on a policy of distributing seeds only to women. Female labour was equally relied on when male labour was unavailable due to migration for work elsewhere. For example, approximately 80 per cent of the cotton cultivated in the zone belonging to the Companhia de Algodão Sul do Save was done by women because most of the men in southern Mozambique were working as miners in South Africa.[48] A report in *Boletim Geral das Colónias* in 1943 also stated that of the 91,236 African producers in Angola, 30 per cent were women and children.[49] Algodoeira Colonial Agrícola in Angola claimed that for its 1949 cultivation it was using women and children in addition to 3,650 men.[50]

Even using women, labour shortages persisted and those companies which had taken out concessions just to protect their male (or joint) labour force sometimes clashed with those companies solely devoted to cotton production. The clashes were particularly frequent during World War II when prices for raw materials had jumped and most companies wished to profit from the increases. Sena Sugar Estates clashed with the cotton concession Lopes e Irmão over the use of labour in the region of Zambezia, but since official interest gave priority to cotton at the time, Sena Sugar Estates had to find alternative sources of labour.[51] Similarly, in the 1950s when Sena thought of dropping its concession for cotton it decided not to because the very people who were interested in it were their old rivals, Lopes e Irmão and Sena worried about losing labour to this cotton concession.[52]

Caught up in the conflict over labour and its use was the colonial administration. Many civil administrators complained that they had too much to do and spent too much time forcing people to work. In addition to forcing Africans to fulfil their cotton obligations, administrators had to assist other companies with labour problems, collect taxes, and ensure that all eligible males did six months work a year for the state (including road-building and other public works).

[48] JEAC, *Alguns anos de actividade* (Lisbon: Cristovão Augusto Rodrigues, Lda., 1946), 30.

[49] 'Algodão de Angola', *BGC* 21: 236 (Feb. 1945), 304–6.

[50] Algodoeira Colonial Agrícola, *Relatório*, 1948, no page.

[51] Head, 'Sena Sugar Estates', 65.

[52] Vail and White, 'Tawani Machambero!', 254–5.

The job of administrator necessarily involved conflicting loyalties and priorities because a number of sources required labour. The 1928 Labour Code had stipulated that administrators were to ensure that labour was treated properly. Proper treatment would, it was supposed, slow emigration and increase tax revenues for the government. The duty to protect that labour, however, conflicted with later legislation on the administrator's duty to aid companies with labour problems. As demands for labour by companies increased during the war, administrators moved from a position of mediating between Africans and companies to one of mediating between companies over labour supplies. In Mozambique especially, as the power of the JEAC grew and demands for cotton by the metropolitan textile industry increased, more often than not priority was given to the cotton campaign. In 1940 the JEAC in Mozambique tried to obtain bonuses for administrators who achieved the best results for cotton production in their zone—a system which had been adopted in Angola to improve cotton production. When Mozambique's Governor blocked the scheme, local cotton companies privately adopted the practice of giving bonuses to administrators.[53] In both 1938 and 1942 an official of the Zambezia Company claimed that local administrators had prohibited him from recruiting in cotton zones.[54] In 1944 the administrator of Lugella in the district of Quelimane, Mozambique was reprimanded by the Governor of Quelimane for not sending workers to Namagoa Sisal Company but the administrator claimed that he was 'obliged to give priority to the cotton campaign'.[55] The administration intervened often, using the African police to enforce the scheme and using chiefs to round up labour.[56]

Concessionary companies were allowed to hire managers or overseers who worked with the administration's *cypaes* and the technical workers of the JEAC to supervise cotton growing. Cotonang in Angola mentioned that its employees were 'working with Africans and spreading propaganda'. In fact, since overseers were often paid bonuses for high cotton output, they frequently abused Africans to get them to produce. A service order of 1941 in

[53] Cahen, 'Corporatisme et colonialisme', Part I, 409.

[54] Vail and White, *Capitalism and Colonialism*, 277.

[55] Ibid. 197.

[56] Isaacman *et al.*, 'Cotton is the Mother of Poverty', 589; and Isaacman, 'Chiefs, Rural Differentiation', 22–9.

Mozambique unknowingly exposed the problem when it prohibited the ill treatment of female cotton growers by overseers—testimony to what had been happening.[57]

Increased state surveillance alongside the vigilant efforts of the JEAC and the concessionary companies had dramatic effects on the number of producers in Mozambique which rose from 534,000 in 1939 to 791,000 in 1944.[58] Though official statistics show a decline in the number of producers in Mozambique to settle at a round half a million from 1950 onwards, Allen Isaacman reckons that these figures do not include a significant number of child labourers who were involved in the scheme.[59] In Angola, the number of producers also rose from 65,987 in 1940 to 80,000 in 1945.[60]

In addition to local administrators who were involved with the cotton scheme on an almost daily basis, the JEAC also intervened and it had no conflict over priorities. It was concerned not simply to improve techniques as its official pamphlets stated but to get as many Africans as possible to cultivate cotton. The bonuses it obtained for administrators and its conflicts with other companies and other administrators reflected this effort.

Cotton company reports repeatedly praise the JEAC for helping them to acquire labour. Cotonang boasted that it had been working with the authorities to increase the number of cultivators in its zones.[61] The JEAC's own annual reports testify to the level of involvement, from recruiting labour to planning settled cotton communities. In 1946 it called attention to its involvement in recruiting (and by inference, to its recruiting methods) when it issued a vehement denial of alleged abuse of Africans. It was true, the JEAC said, that some Africans had worked hard and only produced a few kilos of cotton for their efforts but others had taken hundreds of kilos to market. It was simply not fair that

> Even today, determined critics, not being able to refute the material activity of the Junta, try to diminish its national prominence by alleging that the Junta has abandoned all social concerns.[62]

[57] Vail and White, *Capitalism and Colonialism*, 316–17.
[58] Isaacman *et al.*, 'Cotton is the Mother of Poverty', 588.
[59] Ibid.
[60] *BGC* 26: 300 (June 1950), 195.
[61] See Companhia Geral dos Algodões de Angola, *Relatório* for 1948, 1950, 1954 (hereinafter Cotonang, *Relatório*).
[62] JEAC, *Alguns anos*, 27.

Whether the JEAC was socially concerned or not, the fact remains that it was constantly preoccupied with the dual problems of securing labour for cotton concessionary companies and producing more and better cotton. But these two problems contradicted each other. The JEAC was supposed to improve the quantity and quality of cotton production in Portugal's African colonies. The obvious solution would have been to offer higher prices for unginned cotton, to increase the differences between first and second quality cotton, to encourage Africans to take more care with their cotton harvests, to make sure that unseeded cotton was classified fairly and accurately, and to treat African labour in the cotton zones better than it was treated elsewhere. As it was pointed out, the JEAC did none of these things. The reason was because the JEAC's role in improving the quantity and quality of production for the metropole conflicted with its interest in defending the demands of cotton concessionary companies. Organized to regulate colonial cotton production, instead the JEAC became a defender of cotton company interests in the face of the colonial administration, other companies, producers, and later, even the metropolitan textile industry.

To conclude the period from 1936 to 1946, although production in both colonies had increased, inefficiency and abuse riddled the system. Rather than through high productivity or cost-efficient agricultural practices, increases had been achieved through coercion, punishment, and extensive administrative involvement, and accompanied by resistance, starvation, and conflicts over labour. In their insatiable quest for wartime profits, concessionary companies, aided by the JEAC, sacrificed Africans to the demands of cotton production with predictable results: output was up but cotton quality was poor, quantity was erratic, resistance was rife. With the textile industry poised to launch into foreign markets at the end of World War II, significant changes would have to be effected in the colonial cotton system to satisfy the demands of industry in the post-war period.

7

Pretensions of Democracy and Development, 1946–1958

INTRODUCTION

The Allied victory and the appearance of diverse opposition groups threatened the political and ideological foundations on which the *Estado Novo* rested. After the war, Monarchists as well as corporatists, colonial as well as industrial interests challenged the regime to develop policies coincident with their preferences. The alteration of the Constitution, the creation of corporations, and an anti-Communist stance constituted the multifaceted political strategy the state devised to appease its critics following the war. Most importantly, in a change of style, the *Estado Novo* adopted a technocratic-developmentalist language to deflect political criticism of its authoritarian aspects, a language which simultaneously addressed the changing balance of forces in the Portuguese economy.

From 1946 to 1958 industrial and financial conglomerates which had emerged from the war steered the state towards industrialization and development plans. Portuguese economic policy had a double orientation: to substitute for imports and to become more competitive in traditional export areas. To accomplish these two objectives, the regime geared economic strategy towards the creation of infrastructure and heavy industry, modernization and rationalization. These activities benefited powerful industrial and financial cartels who, in alliance with the state, were entrusted with many of the import substitution plans. Through personal favours, infrastructural improvements, financial contributions, and legislation for the conglomerates, the state also continued to intervene actively in the economy.

The regime simultaneously trumpeted modernization and rationalization plans to allow its traditional export areas to become more competitive. The half-hearted enforcement of these policies, the regime's fiscal conservatism, and the maintenance of protectionism produced an apathetic response by many industries and undermined the strategy. Moreover, traditional interests, such as the textile industry, for the most part were not part of the monopoly capital configuration and their political and economic power began to wane during this period. Less important industries and smaller firms were not wholly ignored nor were they powerless, it was simply that they were not the favoured child.

The configurations devised by the regime proved fragile. By 1958 political and economic challenges resurfaced and the state urgently needed to reassess its policies. It had liberalized enough to gain international acceptance and to divide temporarily the opposition but had not sufficiently altered the degree or nature of its intervention. The means had changed; the objectives had remained the same.

CHALLENGES TO AUTHORITARIANISM

While the government rode the crest of the authoritarian wave in the early years of the war, the increasing likelihood of an Allied victory saw Salazar catering more to Anglo-American demands. In October of 1943 Salazar reluctantly agreed to lend the Azores to the British, rather than have them annexed.[1] The following year, with the Allies near victory, he prohibited the sale of wolfram to the Axis.[2] During the announcement of both of these actions, Salazar judiciously referred to Portugal's long-standing alliance with the United Kingdom.

On the home front, the hardships suffered during the war, organized strikes by workers, and the possibility of an Allied victory gave impetus to the formation of an opposition political movement, the first time a united group had formed to challenge the regime since 1926. The Movimento de Unidade Nacional Anti-Fascista (Munaf), an umbrella organization of Communists, Socialists, Republicans, the democratic left, and anarcho-syndicalists, organized in 1943 and called for the overthrow of the *Estado Novo*,

[1] Salazar, *Discursos*, iv, no. 5, 31–51; and see Tom Gallagher, *Portugal: A Twentieth-Century Interpretation* (Manchester: Manchester Univ. Press, 1983), 102, on British threats to invade in the summer of 1943.

[2] Salazar, *Discursos*, iv, no. 9, 81–3.

though its members divided over the methods by which this was to be accomplished.[3]

By 1945 the period of lucrative neutrality was over as was the *Estado Novo*'s unshakeable command of the political system. Frightened by the Allied victory and their request that Portugal join the 'true democracies',[4] the Portuguese government sought to adapt the old corporatist language to the changed situation, while preserving the essentials of the *Estado Novo*. Salazar tried to use the same words as those of the Allies but attached new meanings to them, describing Portugal in 1945 as an 'organic democracy' and a 'presidential system'.[5] Concurrently, he argued that peace and order were the most important attributes of a nation; the means of their achievement were of secondary importance.[6] In a (meaningless) combination of the old emphasis with the new language, Salazar said on 8 May 1945, 'I do not want to jump to conclusions, but if democracy can have, beyond its political significance, a social significance and meaning, then we are the true democrats.'[7] To demonstrate his apparent conversion, he held a general election and Munaf formed a separate group, Movimento do Unidade Democrática (MUD), to contest it. These elections were not meant to be free, but simply to give Portugal a veneer of respectability with the democratic victors of World War II:

> Elections with opposition candidates who were never elected and with a short period of 'sufficient liberty' (i.e. censorship) before polling day (which could help the regime identify its opponents) were a somewhat farcical ritual, but they did allow the regime a degree of international acceptability; Portugal did have elections, even if the government's candidates did always win. Elections were '*para inglês ver*'; but then Salazar was on record as believing that 'in politics, what seems to be, is'.[8]

By late 1945 Salazar considered constitutional changes to confirm that the regime was not authoritarian and to placate growing criticism. While parties were still banned, the regime considered changes in the representation of the National Assembly and the Corporative

[3] D. Raby, 'Portugal, 1942–1947: O problema da estratégia revolucionária da oposição', *Análise Social*, 20: 84 (1984–5), 692.
[4] Salazar, *Discursos*, iv, no. 12, 101–22.
[5] Ibid. iv, no. 18.
[6] Ibid. iv, no. 12.
[7] Ibid.
[8] Robinson, *Contemporary Portugal*, 68.

Chamber and amnestied 300 political prisoners.[9] Any political liberalization, though, was meant mainly for international consumption: Portugal was trying to get into the newly established United Nations. In reality, the regime reduced the powers of the National Assembly, while in the same year the regime reorganized, renamed, and strengthened the state secret police and gave the Ministry of Justice more arbitrary power.[10] In spite of its propaganda efforts, the regime's tactics fooled no one, not even international observers. In 1946 the United Nations denied entry to Portugal, an issue on which opposition groups concentrated.[11]

In a move again designed to win international approval rather than to satisfy internal opponents, the government announced a presidential election in 1949. Many of the opposition groups, comprising both left and right elements (even many Monarchists), united in support of the campaign of General Norton de Matos. De Matos had been a former Governor-General and High Commissioner of Angola during the Republican period and had an undeserved reputation in the popular imagination as having been successful.[12] His credentials as a loyal Republican and an opponent of the regime, however, were beyond dispute.[13] During the campaign he demanded that the regime establish and respect basic democratic principles and fundamental liberties. Moreover, he returned to his favourite theme of colonial autonomy, meaning not legal independence (which was on the agenda of few opposition members at this time), but settler control. These were sweet words to the several colonial settler groups who, while frightened of independence movements occurring in other African countries, were nevertheless frustrated with the sacrifice of the colonies to the metropole and the lack of industrialization. Willingly, they threw in their support for de Matos.

To the campaign, the government responded with censorship, harassment, and even imprisonment of opposition members.

[9] Cruz, *Monárquicos e republicanos*, 105.

[10] Manuel Braga da Cruz, *O Partido e o Estado no Salazarismo* (Lisbon: ed. Presença, 1988), 90–1.

[11] Robinson, *Contemporary Portugal*, 69.

[12] Clarence-Smith, *The third Portuguese empire*, see ch. 5 on the activities of de Matos as High Commissioner in Angola during the Republic.

[13] D. L. Raby, *Fascism and Resistance in Portugal: Communists, Liberals and Military Dissidents in the Opposition to Salazar, 1941–1974* (Manchester: Manchester Univ. Press, 1988), 29–33.

Despite the fact that General de Matos withdrew the day before the election, the activity and strength of his campaign shocked the regime. In response, Salazar proceeded with a four-pronged attack designed to achieve overlapping international and national objectives. The four areas addressed were constitutional revision, the creation of corporations, anti-Communism, and the adoption of a technocratic-developmentalist language.

First, Salazar acted on his earlier thoughts and revised the Portuguese Constitution in 1951. In it he attempted to mollify metropolitan and colonial opposition and international criticism by asserting the permanence of the system and highlighting its 'democratic aspects'. Furthermore, arguing that 'it is intrinsic in the Portuguese nation to fulfil its historic mission of colonization in the lands of the Discoveries', the Constitution responded to what Salazar euphemistically called the 'crisis in colonial thought' or rather, the post-war calls for self-determination in African countries, by politically integrating Mozambique and Angola with Portugal.[14] They became 'provinces' of Portugal thus notionally enjoying the same legal status as Alto Alentejo, for example. The Portuguese government later took refuge in these semantic changes when confronting charges that it still maintained colonies. But for the present the constitutional claim that 'the Overseas Provinces, as an integral part of the Portuguese state, are united as between themselves and with Metropolitan Portugal', placated critics at home and abroad: Portugal was admitted to the United Nations in 1955.[15]

The 1956 creation of corporations demonstrated further the pressure that was being applied against the regime and how far the regime was willing to go to reduce it. Howard Wiarda views the creation of corporations in the late 1950s as an indication of a 'corporate revival' in Portugal;[16] it is more plausible to regard their creation as a symbolic exercise designed to meet external criticism with polemics and to pacify traditional conservative supporters of the regime, such as the Monarchists.

The 1951 constitutional revision coincided with the death of

[14] 'The Portuguese Constitution', in Amos J. Peaslee, *Constitutions of Nations: Europe*, iii (The Hague: Martinus Nijhoff, 1968), ch. 7, sect. 1, Art. 133.

[15] Ibid. Art. 135.

[16] Wiarda, *Corporatism and Development*, see discussion on pp. 183–208.

President Carmona and gave Monarchists in the regime an opportunity to return to the major tenet in their platform—the restoration of the Monarchy.[17] The Monarchists had been major collaborators with the regime but when Salazar once again rejected their claim, they turned against him. Monarchists withdrew in protest from the National Assembly elections in 1953 and boycotted the National Assembly and the Corporative Chamber in 1955. By the late 1950s, more Monarchists had joined the opposition, and governmental relations with this group were strained.[18]

The creation of corporations was an attempt to appease these Monarchists and other sectors of the opposition within Portugal, or to counter their criticisms. To the Monarchists it showed that the regime had a permanent and institutionalized character and was not incomplete as they had charged. To the corporatists, it suggested that the regime was moving ahead with promises it had made in the 1930s. To international observers and democratic elements within Portugal, it appeared to decentralize power away from the state and into the hands of notionally, non-governmental organizations. The regime created several corporations from 1956 to 1958, but there is no evidence to indicate they had either a well-defined role or any real power. 'Decision-making authority was never actually transferred from the powerful state apparatus, and public policy-making was limited to the bureaucratic arena.'[19] The state intended to remain firmly in charge and interventionist.

Salazar and his supporters took every opportunity to polemicize about the merits of corporatism in general and corporations in particular. Not only did the *Estado Novo* try to convince the world that it did support a form of representation and was not a dictatorship as had been suggested by other members of the United Nations, but it counterbalanced the virtues of corporatism neatly against the vices of Communist governments, a theme frequently touted by the regime during the 1950s and one designed to further weaken the Portuguese left, which was split over the Cold War.[20] Aware of the contradictions between his support of the Allies in the final years of the war and his own authoritarian regime, Salazar now tried to secure legitimacy by joining in the Cold War and the

[17] Cruz, *Monárquicos e republicanos*, 197.
[18] Ibid. 202–9.
[19] Opello, *Portugal's Political Development*, 58.
[20] Raby, *Fascism and Resistance*, 33.

attack on Communism. Domestically, the Communist threat was used not only to encourage greater unity within Portuguese society and to divide and discredit the opposition (the democrats from the Communists) but also as another justification for the maintenance of corporatism. In an interview with the foreign press in 1958, the Minister of Corporations and Social Welfare declared:

> We are corporatists because we are convinced that man cannot be dealt with as an abstract being separated from the family to which he belongs, the occupation or profession in which he works, and the local community in which he lives. We think that it is precisely in this real sphere of his interests that he can take an active part in the life of the Nation and State. We would go further and say that we believe that it is through autonomous corporative institutions that he will best be able to defend himself against the statist tendencies that are so prevalent today.[21]

These 'statist' tendencies to which the Minister of Corporations referred were of course directed at Communist governments. From a defence against the party divisions of liberal democracy in the 1930s, the pretense of Portuguese corporatism became a bulwark against statist trends in the 1950s.[22]

Finally, supporters of the *Estado Novo* engaged in diversionary tactics to shift the focus away from the nature of the regime to its accomplishments. The regime adopted a technocratic-developmentalist language, and drew up development plans, a strategy which Franco's Spain also adopted during the 1950s. Although the language and the plans addressed the changed economic circumstances in the country and the new constituency which had grown along with it, they also served conveniently to direct attention away from politically sensitive topics such as democracy, representation, and rights, and towards the safer ground of production targets and growth.[23] At the same time, the image of a developing, modernizing, and industrializing Portugal dovetailed nicely with the old nationalistic message. Development was portrayed as the modern road to national independence and therefore the government's adoption of it did not contradict an earlier emphasis on self-sufficiency. The adoption of a language of development constructed

[21] Henrique Viega da Macedo, *Alguns principios da política social e corporativa portuguesa* (Lisbon: Ministério das Corporações e Previdencia Social, 1958), 89.
[22] Salazar, *Discursos*, v, no. 4, 47–72.
[23] See ibid. iv, no. 37, 481–511; and v, no. 4, 47–72.

and articulated by lawyers and engineers closely associated with the regime distinguishes this period ideologically from those preceding it, and in the discussion of the economy which follows we will seek to discern how closely the reality corresponded to the propaganda.[24]

POST-WAR ECONOMIC STRATEGY

Economically, the government's adoption of development and modernization plans addressed the changed conditions brought by war: the growth of financial-industrial conglomerates, the expansion of exports, the substitution of imports, and the social hardships endured due to wartime shortages. The developmentalist orientation of the economy was a *de facto* recognition by the state that war had derailed the corporatist project of a self-reliant, self-sufficient economy and had demanded social sacrifices which must be rewarded in the post-war period. Not only must Portugal be prepared through industrial reorganization, modernization, and expansion to participate in a more interdependent, competitive post-war world (while of course preserving its political independence) but it must allow Portuguese labour, on which so much of the war's burden had fallen, to share in any growth through higher wages, lower costs, a higher standard of living, and increased consumption.

The general lines of the development were first traced in the Law of Industrial Development and Reorganization (Lei de Fomento e Reorganização Industrial), the brainchild of Ferreira Dias, the pro-industrialization Under-Secretary of Commerce and Industry from 1940 to 1944, who would return as Minister of the Economy for four more years in 1958. The Law of Industrial Development and Reorganization was passed in 1945 but implemented after the war. It recognized the enlargement of the industrial base which had occurred and committed the regime to direct and indirect financial participation in certain industrial activities. The government provided a budget of 1.5 million contos for the creation and development of iron and copper production, nitrate and cellulose

[24] José Maria Brandão de Brito, *A industrialização portuguesa no pós-guerra (1948–1965): O Condicionamento Industrial* (Lisbon: Publicações Dom Quixote, 1989), 83.

fertilizers, electrification and energy projects, and the improvement of existing industries in order to substitute for costly imports. In traditional sectors, the law anticipated the elimination of small, unproductive firms and their replacement by large, modern, efficient factories in order to improve their abilities to compete abroad and domestically.[25]

What were the objectives of this law? First of all, it acknowledged that industry and finance, rather than agriculture and commerce, were coming to dominate Portuguese society. Secondly, within industry and finance, it recognized the emergence of groups of centralized and concentrated capital as a result of state actions since the 1930s and the advent of war from 1939 to 1945. Thirdly, its provisions sought to aid these groups economically by offering state support through the expansion of infrastructure or direct investment in industrial projects. According to Elisabeth Leeds, large capitalists immediately took advantage of the law which 'helped to consolidate Portugal's *grupos económicos* or banking and industrial monopolies'.[26]

In assigning a prominent role to the government, the law also served to reinforce the government's position in the economy and cement its relationship with finance and industry. Through the supply of credit and share purchase, the state hoped to gain support from large capitalists, control foreign investment, and retain its interventionist approach. With regard to the traditional industries with more dispersed ownership, corporate organisms and industrial regulations remained the preferred method of state intervention. Additionally, industrialization which aimed to replace imports would contribute to a healthy balance of payments and also reduce Portugal's reliance on the outside world. Reduced reliance translated into decreased international interference in Portuguese affairs, one of Salazar's enduring obsessions. Finally, the law recognized that the war had led to a deterioration of industry, and subsequently, stagnation in both productivity and wages. Supporters of the law explained that modernization would bring about an increase in productivity which would translate into higher

[25] Elizabeth Leeds, 'Salazar's "Modelo Economico": The Consequences of Planned Constraint', in Thomas Bruneau *et al.*, *Portugal in Development* (Ottawa: Univ. of Ottawa Press, 1984), 26.

[26] Ibid. 26.

wages for workers. In turn, these higher wages would lead to increased consumption which would then benefit import substituting industries.[27] The state even attempted to take an early step in that direction by advocating wage increases in many industries after the war.

Subsequent decree-laws passed in 1947, 1950, 1952, and 1954 reduced the number of industries subjected to industrial regulations and encouraged private initiative.[28] The adoption of a more liberal approach to those sectors in which families or groups were dominant again revealed the regime's willingness to favour the more powerful economic élites and conveyed a recognition that concentration had occurred. Moreover, from 1944 some industries were allowed to locate in the Portuguese colonies. This must not be interpreted to mean that the regime was liberalizing all of the economy, but rather, as another effort to aid emerging financial and industrial interests, to gain political support, and to tie the colonies more tightly to the Portuguese economy. Further aid was offered through the creation of public credit institutions such as the National Development Fund (Fundo de Fomento Nacional), the availability of Marshall Plan Aid from 1949, state financing of enterprises, and greater connections with other European economies through participation in the OEEC, the European Payments Union, and the European Monetary Accord.

The First Development Plan which followed in 1953 allocated 13.5 million contos to metropolitan and overseas development—half to Portugal and half to all the colonies.[29] For metropolitan Portugal, the sum was subsequently raised to 10 million contos in 1955 and 11.5 million contos in 1957. In the metropole, the plan directed credit, investment, and loans towards the powerful families and industrial groups which had grown up or expanded during the war. Industry and communications received the bulk of development spending with special attention paid to the development of hydroelectricity, petroleum-refining, communications improvements, and iron and steel metallurgy.[30]

In the colonies, the plan was geared towards internal develop-

[27] Ferreira do Amaral, *Os novos caminhos da indústria portuguesa* (Lisbon: Min. de Economia, 1947), 16–17.

[28] Cruz, 'Política industrial', 12–14.

[29] Salazar, *Discursos*, v, no. 7, 97.

[30] *BGU* 28: 326–7 (Aug.–Sept. 1952), 10.

ment though the government allocated scant resources for health, education, or social services. The initial allocation of 7 million contos in the First Development Plan was actually cut to 4.875 million contos from 1953 to 1958.[31] The remainder aimed at general improvements in infrastructure, communications, oil and mineral exploration as well as the creation of planned settler communities to attract emigrants.[32] Emigration became conscious state policy at this time not only because the state wished to strengthen its presence in the colonies but more importantly, because it wanted a safety net for its development policy at home. In the event that growth did not proceed as smoothly as planned, the emigration could serve a twofold purpose. First, by attracting labour and thereby reducing the supply at home, the colonies acted as an indirect pressure on metropolitan industry to raise wages. Secondly, by absorbing unemployed or excess labour, the colonies mitigated the effects of modernization.[33]

The *Estado Novo* additionally hoped to provide an attractive and protected area of investment in the colonies for Portuguese business, especially financial-industrial conglomerates, and to mollify settler grievances by fostering industrialization, colonization, and mechanization in the colonies. With state support, conglomerates and settlers could take the opportunity to invest capital accumulated during the war (most foreign investment was controlled after 1944 until the 1960s) which would further unify the colonial and Portuguese economies.

If the state was prepared to adopt development in this period, nevertheless it did so according to strict criteria. Foremost among these were financial and social stability. From 1946 to 1958, in spite of the stated aims of its development plans, the Salazar Government continued to protect and favour Portuguese industry and agriculture by maintaining or ignoring low wages, placing high customs duties on imports which competed with domestic production, and enforcing industrial regulations to limit competition. During the 1950s, wages rose only slightly and price increases were negligible.

[31] Portugal, Assembleia Nacional, *II Plano de Fomento (1959–1964)* (3 vols.; Lisbon: Imprensa Nacional, 1959), i. 5.

[32] *BGU* 28: 326–7 (Aug.–Sept. 1952), 11–17.

[33] For a thorough examination of labour export policies, see Elizabeth Rachel Leeds, 'Labor Export, Development, and the State: The Political Economy of Portuguese Emigration', Ph.D. thesis (MIT, 1984).

Additionally, mindful of the Republicans' spending spree after World War I and the financial and social instability that followed, the Salazar Government exercised rigorous control over the balance of payments. The war had contributed to a considerable build-up of foreign-exchange reserves which had been helped by three years of positive trade balances. But by 1948 the balance of payments was again negative due to increased imports and decreased exports. Having re-encountered the negative balance of payments, the *Estado Novo* resorted to old remedies: the use of emigrants' remittances to prop up the deficit, import controls, and a reinforcement of the links with the colonies.

As so many Portuguese governments had done in the past, the *Estado Novo* used emigration to relieve its balance of payments' woes. Emigration increased from 8,275 people in 1946 to 12,838 people in 1947. By 1958 an average of 30,000 people a year were emigrating.[34] These offset but did not eliminate the balance of payments deficit; the government then resorted to strict foreign-exchange control. Throughout the 1950s, the escudo to sterling and escudo to dollar convertibility rate remained virtually unchanged (Tables A1 and A2). The regime also repeatedly stated that borrowing abroad would be limited and growth controlled. In his discussion of the First Development Plan, Salazar said:

> While a quarter century of well equilibrated finances and balanced accounts could justify some boldness, administrative consistency and financial stability continue to be the best base of national reconstruction. Furthermore, the application of our own resources—as has been seen—is the healthiest measure for the reinforcement of our political independence. Too much pressure on the domestic economy . . . works against the monetary stability and social equilibrium which we have been resolutely defending. Equally, due to the uncertainty and precariousness of world conditions, neither would the excessive resort to external credit safeguard our best interests.[35]

Instrumental in safeguarding their best interests and easing the pressure on the balance of trade would be the colonies. The 'moral and economic unity' cemented by the new Constitution indicated that the *Estado Novo* was determined that the colonies continue to serve the metropole, although in a more expanded capacity than the straightforward mercantilism of the past. Still, the Salazar regime

[34] Serrão, *A emigração*, 31–2.
[35] Salazar, *Discursos*, v, no. 7, 95.

expropriated the consistent positive balances of payment which the colonies had with foreign countries throughout the period to reverse Portugal's own negative balance of payments. For example, Portugal's negative balance of payments of 931,000 contos in 1950 became a positive balance of payments of 471,000 contos owing to the addition of the colonies' surplus with foreign countries; and Portugal's 1958 deficit of 2,189,000 contos became a surplus of 825,000 contos due again to the colonial contribution to the total balance of payments.[36] While negative trade balances remained a permanent feature of the balance of payments, these were offset by remittances, capital transfers, and the colonies' positive trade balances with foreign countries.

RESULTS OF POST-WAR POLICIES

Within this environment of stability and control, the government honoured its commitment to infrastructural improvements during the 1950s. Chief among these improvements was the development of hydroelectricity as outlined in the National Electrification Law of 1944 and implemented after the war. Financed with more than 60 per cent public funds and the rest from the private sector and external credit, the construction of hydroelectric dams and reservoirs led to an increase in energy output from 940,000 kilowatt/hours in 1950 to 3,260,000 kilowatt/hours in 1960.[37] SACOR, supported with one-third state capital, also increased its oil-refining capacity from 480,000 tons per year to 1,200,000 tons.[38] Additionally, roads were expanded and modernized. As a result registered commercial vehicles rose to 48,700 in 1959 as compared with 26,000 in 1949, and the number of private vehicles from 56,500 in 1949 to 145,000 in 1959. Port facilities were improved in Lisbon and Oporto, and airports constructed in those two cities. The state also financed the creation of the Portuguese national airline (TAP) in 1946.[39]

[36] Edgar Rocha, 'Colónias e exportação de mão-de-obra como fontes de divisas: Considerações sobre a contribuição dos emigrantes para o subdesenvolvimento económico português', *Análise Social*, 18: 72–4 (1982–3), 1059.

[37] Robinson, *Contemporary Portugal*, 138; and J. M. Felix Ribeiro *et al.*, 'Grande indústria, banca e grupos financeiros (1953–73)', *Análise Social*, 23: 99 (1987), 954.

[38] Ibid. 957.

[39] Robinson, *Contemporary Portugal*, 137.

These improvements together with the assistance granted by the Law of Industrial Development and Reorganization and the First Development Plan led to some growth. Portugal's GNP increased by 4.1 per cent between 1951 and 1957, largely due to the 6.5 per cent growth rate of Portuguese manufactures between 1953 and 1958.[40] The GNP per capita went from US$195 to $250 between 1954 and 1959.[41] Gross fixed investment as a percentage of Gross Domestic Product was 14.5 per cent from 1952 to 1956. Additionally, in the colonies, investment by private and public capital in railroads, ports, infrastructure, communications, oil and mineral exploration, and agriculture also took place.

For the modern industries in particular, the development plans, trade agreements, and state financial and legal support contributed to growth and aided the continued formation of large conglomerates by the early 1960s. The index of industrial production for modern or non-traditional industries such as cement, paper, chemicals, metallurgy, transport equipment, and plastics was 246 in 1960 compared with the base of 100 in 1953.[42]

Moreover, amongst the modern or non-traditional industries and those with monopolies in their field after World War II, ownership was consolidated and share purchase in other companies expanded. In Chapter 4, we saw that powerful groups formed and expanded in the period from 1936 to 1946 not only because of wartime trade and production opportunities, but also because the regime either turned a blind eye, granted special concessions, or used legislation and organizations of economic co-ordination to help the consolidation of monopolies. During the 1950s, similar policies prevailed as the government adopted almost a *laissez-faire* attitude towards big business.

Most of these conglomerates had an intimate relationship with the government. Most operated in sectors devoid of formal primary or secondary corporate control in the form of *grémios*, commissions, institutes, or *juntas*. The *Estado Novo* meant *juntas* for some, ministries for others:

industrial elites, especially in Lisboa, especially Companhia Uniao Fabril (CUF), the Champalimaud group, and the Português do Atlantico group

40 *EFTA Bulletin*, 10: 6 (July–Aug. 1969), 13.
41 *EFTA Bulletin*, 9: 8 (Nov. 1968), 4–18.
42 Pintado, *Structure and Growth*, 173.

were able to pursue their interests via direct contact and alliances with different sectors of the state apparatus, including the military.[43]

In a 1965 study, almost half the industrial élite surveyed had held a public or corporate position; of the 26 per cent who held public positions (such as membership in the National Assembly or the Corporative Chamber or office at the municipal level), 'heads of larger enterprises were more likely to have occupied public positions at the national level of the government'.[44] Those who served at the national level tended to be from the more modern sectors in the Lisbon–Setúbal area, both state and private, and to have technological backgrounds. While it was rare for a top executive to move from industry into the government at the cabinet level, it was not unusual for government officials to move into industrial positions.[45] This unofficial and official contact between government and industrial élites—so extensive as to merit the description 'kleptocracy'[46]—no doubt influenced the way industrial élites were treated.

Aided by their influential position with government officials, development plans, trade agreements, and low-wage policies, these conglomerates were allowed, if not actively encouraged to expand during the 1950s and 1960s. For example, CUF and Champalimaud benefited enormously from the state's highly protectionist commitment to colonial and domestic development. The 1950s saw CUF expanding into fertilizers, edible oil and fibre processing from Africa, and tobacco processing.[47] Similarly, Champalimaud had a monopoly on the production of cement in both the colonies and the metropole by the early 1950s. By 1954 the Champalimaud group was directing the construction of a national steel industry in which it participated as a major shareholder along with other public and private capital.[48] In 1955 the company was granted a ten-year monopoly with the possibility of an extension granted in 1957.[49]

[43] Harry M. Makler, 'The Portuguese Industrial Elite and its Corporative Relations: A Study of Compartmentalization in an Authoritarian Regime', in Graham and Makler (eds.), *Contemporary Portugal*, 127.

[44] Ibid. 129.

[45] Ibid. 137–40.

[46] Herminio Martins, 'Portugal', in M. S. Archer and Salvador Giner (eds.), *Contemporary Europe: Class, Status and Power* (London: Weidenfeld and Nicolson, 1971), 69–70.

[47] Ribeiro *et al.*, 'Grande indústria', 949.

[48] Ibid. 961. [49] Ibid. 962.

Yet the late 1950s found few, including the large conglomerates, particularly satisfied with the pace of development since 1946. In reality, the Law of Industrial Development and Reorganization and the First Development Plan were extremely limited in their approach to overall development and disappointing in their results. In comparison with other countries, Portugal's economy looked sluggish. Compared with a growth rate of 4.1 per cent in Portugal, the rate of growth in Greece was 6.9 per cent and in Turkey 5 per cent, even though both countries had rates of investment very similar to those in Portugal.[50] Moreover, despite the emphasis on improving exports from Portugal, their growth was only 12 per cent between 1954 and 1959.[51] In addition, the export structure had changed little and was still dominated by traditional exports.

While financial-industrial conglomerates had benefited from state policies and support, stagnation and deterioration characterized the rest of the Portuguese economy in older, traditional industries as well as agriculture. Agriculture was backward, inefficient, with low yields; it needed higher prices to modernize or government investment, and reform of its structure. Traditional industries were also in crisis because many had not modernized and could not compete abroad. With 1953 as the base year, the index of industrial growth for the cork, leather, textiles, food processing, beverages, and resin transformation sectors was only 136 in 1960 as compared with 246 amongst non-traditional industries cited earlier.[52] Others experienced growth but without a concomitant growth in trade. For these, obstacles to trade still existed. Increasing exports was difficult because the government had concluded few export agreements with foreign markets; selling on the domestic market was limited because of the low purchasing power of consumers. Since neither agriculture nor the traditional industries had modernized or expanded, wages had not increased.

A number of reasons explain why little change occurred in most of the Portuguese economy during the period. In traditional sectors, the government was still highly interventionist. It maintained a strong grip on these areas through its corporate organisms and laws on the regulation of industry. Industrial Regulation for

[50] Pintado, *Structure and Growth*, 19.
[51] *EFTA Bulletin*, 10: 1 (Jan. 1969), 101.
[52] Pintado, *Structure and Growth*, 173.

these industries remained in force even after a supposed relaxation of the terms of the decree in the 1950s placed greater emphasis on private initiative. According to one critic, the legislation operated as before, preventing the creation of new companies and requiring government authorization for the acquisition of new machinery whether for replacement or expansion.[53] Between 1948 and 1965, 75 per cent of the requests for equipment, expansion, or the establishment of new firms related to food-processing, beverages, textiles, clothing, tobacco, and leather—all traditional industries.[54] The state continued to deny authorization for the creation of new firms in these sectors (60 per cent of the applications for new ventures were denied) yet little restructuring took place, in spite of the fact that the state granted around 70 per cent of all requests for the replacement or addition of machinery.[55]

Among the traditional industries of cork, wine, resins, and textiles, although Law 2,005 of 1945 had sought the elimination of small factories and the creation of large, efficient ones, its implementation was arbitrary. The government did support some modernization in order to satisfy powerful industrial economic groups within these sectors, to increase exports, or to substitute for imports. But for most traditional industries as well as agriculture, state policy was not nearly as consistent nor accommodating as in the modern sector. Older, dilapidated companies survived and modernization only took place rather haphazardly among the more powerful companies. Since growth meant higher imports of technology and machinery which affected the balance of payments, modernization in these industries was pursued as long as it did not conflict with fiscal stability. While the government did not actively shut factories, neither did it offer sufficient long-term credit or other incentives.

If the state still extensively intervened in these sectors and cautiously approached modernization, it is also true that many firms in the traditional sector manipulated the state. According to a report by the Corporative Chamber, some factories used Industrial Regulation to block requests for new businesses within the regulated sectors, and thus to prevent growth and competition.[56] Many firms

[53] Cruz, 'Política industrial', 14.
[54] Brito, *A industrialização*, 275.
[55] Ibid. 274.
[56] Portugal, Assembleia Nacional, *II Plano*, ii. 126–7.

also obstructed government attempts to enact even minor wage increases, which would have forced them to modernize. Additionally, small firms were able to twist the policies to their advantage by evading Industrial Regulation laws, using family labour, working in illegal conditions, and dumping clandestinely on the domestic market (see Ch. 8). Fears of social disruption may explain the government's reluctance to push ahead with far-reaching development plans which might have lead to factory closures and unemployment. In spite of its much vaunted emigration policy to the colonies, it was unwilling to eliminate labour-intensive industries. Such action might have jeopardized the delicate social balance that underlay the regime.

Since many traditional industries had not modernized, they relied on low wages to remain competitive and to survive. Unfortunately, stagnant wages led to low domestic consumption making the domestic market too small for a viable import substitution policy. In turn these stagnant wages prompted an increase in social conflict and a subsequent rise in the opposition, and by the late 1950s would lead to emigration to Northern Europe. In an attempt to avoid growth-related, social instability, the state had only provoked another kind of instability—emigration abroad—and ultimately, on a scale it could not control.

Along with the poor statistics and growing social conflict came criticism of the nature and pace of development which had taken place. In spite of the privileges they enjoyed in the Portuguese economy, the financial-industrial oligopolies were now in a position to expand abroad and denounced the state's apathy in promoting Portuguese exports. Alongside an increase in quantity and quality of goods produced, they wanted an increase in markets.[57] Citing the criticisms of many modern factories, even a government body, the Corporative Chamber, called for an improvement in quantity and quality of goods in order to compete in European markets and the modernization and concentration of factories in order to improve economies of scale. With the formation of the European Economic Community (EEC) in 1957, Portuguese industrialists feared isolation and reduced markets if export agreements were not actively pursued.

[57] See esp. the discussion by Carlos Garcia Alves, 'Desenvolvimento industrial e problemas de mercados', in *Relatórios de problemas gerais*, II Congresso da Indústria Portuguesa (Lisbon: AIP, 1957).

The desire to improve exports and to join the new trading configurations in the rest of Europe were just the tip of the iceberg. Imbued with a technocratic ethic, these industrialists and their government supporters wanted less direct intervention but greater government financial participation in the reorganization and modernization of industry, an acceleration of development, and more liberal credit policies. They resented the fiscal conservatism of the state with its restrictions on credit and joint ventures with foreign companies. They demanded liberalization. In its opinion on the Second Development Plan, the Corporative Chamber supported the views of the modern industrialists. It argued that while development was proceeding, the process was too slow. The government was overly cautious and its attitude towards private initiative was hostile.[58]

In the traditional sector, critics called for an end to arbitrary governmental treatment and requested support for modernization. They wanted more credit, lower taxes, and export opportunities. Industrial Regulation also came in for a lot of criticism. One critic, Fernando Cruz, denied any challenge to the system, but he pointed out that Industrial Regulation had been implemented at a time of crisis and that no study had preceded its implementation. Years had gone by, industries had been created or expanded and still no study of its operation had taken place. In effect, the government could not possibly know what it was regulating, or how well.[59] By the mid-1950s, economists, technocrats, and some businessmen sought an end to Industrial Regulation and protectionism in these industries because they felt that their being shielded from competition had eliminated incentives for investment and modernization.

Secondly, it was suggested that some competition might be healthy for the country not only to improve the quality and prices of products but also to challenge the monopolies which had grown up. While the state should continue to 'orient, guide, and stimulate the economy', competition ought also to have a role and regulation reduced.[60] Thirdly, at the Second Industrial Congress (II Congresso da Indústria Portuguesa) in 1957, Portuguese industrialists and economists warned of the urgency of modernizing Portuguese

[58] Portugal, Assembleia Nacional, *II Plano*, 'Pareceres da Camara Corporativa', ii. 261.

[59] Cruz, 'Política industrial', 20.

[60] Ibid. 34.

industry in order to participate in the new European trading associations—the EEC and EFTA.[61]

That the state anticipated or intended these results is unlikely. The post-war period demonstrates that the *Estado Novo* was promoting development, even if cautiously. The period reveals a willingness to finance growth, industrialize further, increase exports, substitute for imports, and in doing so, raise the standard of living. At the same time, however, the regime burdened its industrial strategy with an obsessive commitment to financial and social stability resulting in only weak development confined to monopolies. It did not fully take advantage of the benefits garnered from neutrality during World War II to develop the country and modernize Portuguese industry.

By 1958 these conflicting objectives put the state in a precarious position. The state had not anticipated that its technocratic-developmentalist language would be adopted by industrialists and even government members, and that the financial-industrial conglomerates it had nurtured would want to expand abroad. It did not foresee that its control of traditional industries would reverberate in the wage structure, and domestic and foreign markets. The government's myopia entailed costs. Modern groups were impatient with the state's financial strategy and lobbying for greater links abroad. Traditional sectors chafed against industrial regulations and to their criticisms were joined those of workers outraged at the low salaries. In addition, colonial interests were dissatisfied with the pace of development and argued for greater autonomy. All of these grievances contributed to an upsurge in political opposition to the regime in the period after 1958. To secure its hold on Portuguese society, the government had to redesign its political and economic strategies.

[61] See articles by Alves, 'Desenvolvimento industrial'; and Cruz, 'Política industrial'.

8

Post-War Crisis in the Textile Industry

The reorganization, modernization, and expansion programme which was implemented for the Portuguese textile industry following World War II mirrored the general developmentalist approach that began to be more consistently formulated by the *Estado Novo* in the post-war period; and it recognized some of the negative results of Industrial Regulation. In response to a changed international environment, the wartime economic growth of Portugal, and poor conditions within many textile mills, the *Estado Novo* began to advocate technical solutions to the difficulties of the industry. On the eve of the industry's encounter with modernization, unfortunately, the problems of competition, the cotton supply, and textile markets had reared their ugly heads again. It remained to be seen if the new strategies would successfully convert the industry into a profitable venture.

POST-WAR LEGISLATION

In a previous chapter on textiles, we saw how the textile industry had been unable to replace machinery due to wartime conditions and a strict application of industrial controls. Anticipating the end of the war and the loss of exports, many industrialists argued for the necessity of modernization and expansion to meet post-war challenges. Following the war, the government passed legislation designed to improve the industry so that it could attract and retain export markets. The objectives of this legislation and the way it was administered reveal arbitrary, perverse treatment of the industry by the government. Moreover, old battles over competition, prices, and markets as well as the recent dilemma of confronting a new Europe while tied to the colonies indicate the obstacles faced by the industry in the period up to 1958.

Declaring that its regulations, 'should not present difficulties to the re-equipment of our cotton textile industry, so that it can technically and economically satisfy the public's needs and the requirements of its own development',[1] the government issued a circular in 1946 to correct, once again, structural deficiencies in the industry. The circular was a specific application of the 1945 Industrial Reorganization Law (discussed in Ch. 7). Authorization for new machinery would be given, in principle, if the request fell under the following categories: the replacement of old by new machinery according to the criteria outlined in the 1940 legislation (the substitution of one or two manual looms depending on their size by one mechanical loom); the creation or expansion of new plants *as long as* they had a minimum of 10,000 spindles; and the installation of new spinning mills within weaving plants *as long as* the weaving plants possessed a minimum of 300 mechanical looms. In plants which contained both spinning and weaving, the government would authorize an increase in the number of spindles if there was a minimum of 200 mechanical looms, and approve new spinning mills if 200 automatic looms were going to be used.[2]

The justification for this type of correction for the industry was technical and political. Technically, it was part and parcel of the post-war technocratic, developmentalist rhetoric and approach adopted by the government with regard to the Portuguese economy. Connected to this, it illustrated the government's recognition—which it would extend in the 1950s—that its policies could no longer be Malthusian and protectionist, and that Portugal would eventually have to move into foreign markets as foreign suppliers, would, in turn, move into Portugal. The government argued that with the war's end, other countries would modernize their textile industries and Portugal would need to modernize too in order to compete successfully in external markets. The preamble to the 1946 circular argued that the 1940 law had not sufficiently corrected the poor state of the industry.[3]

In the years immediately following the war, domestic competition had already resurfaced. Several companies complained in their annual reports of the highly competitive situation and consequently

[1] *Bol. da DGI*, no. 485 (25 Dec. 1946), 'Despacho normativo' (17 Dec. 1946).
[2] Ibid.
[3] Ibid.

of falling prices in the internal market. As one spokesman for the industry stated:

> The industry regrets, especially, the great drop in internal consumption and the existence of large stocks in the warehouses, which surely provokes uncontrolled competition among manufacturers with a subsequent drop in prices.[4]

Moreover, prices were dropping in the export market entailing a reduction in profits. After the lucrative years of the war, profits went from 28,414 contos in 1946 to 22,349 contos in 1949 even though the quantity of exports abroad increased.[5] Manufacturers were thus forced to rely on the limited colonial market, of which they took full advantage. Settlers complained that the metropolitan textile industry sold goods to the colonies at inflated prices. A comparison of the costs per kilogram of Portuguese textiles exported abroad with those exported to the colonies confirms this criticism. Except for 1951–2, the price of Portuguese textiles exported to the colonies was consistently higher than that of textiles exported to foreign markets. In 1949, for example, the cost per kilogram of Portuguese cloth sold to the colonies equalled 67$41 escudos compared to the cost of 64$08 escudos per kilogram of cloth exported abroad—a difference of 3$33 escudos. In 1953 the difference between the colonial price and the foreign price was 7$39 escudos.[6] While the differences in average cost per kilogram may be accounted for by a difference in quality, this is unlikely, since Portuguese textile manufacturers sold coarser cloth to Africa and finer cloth abroad. To compensate for the failure to improve the return on exports abroad, textile industrialists milked the colonies.

Modernization, the government argued, would help the industry capture more markets abroad. A decrease in the number of factories would reduce domestic competition. Under the terms of the 1946 circular, the emphasis was on replacing old by new machinery and on establishing a certain minimum size of plant: a minimum that was very large in comparison with the average for the majority of firms at that time. None the less, the government justified the transformation as the only way to produce more economically and therefore, to be more competitive:

[4] Carlos Bastos, *Comércio e indústria têxtil em Portugal* (Oporto: GNIAR, 1950), 162.

[5] Ibid. [6] Bastos, *Indústria e arte têxtil*, Appendix, p. vi.

It seems that in this case cutthroat competition can only be successfully overcome by large industrial organisms, adequately equipped with capital and machinery. By operating at the lowest possible rate of profit these industries will drive out small industries by capturing markets.[7]

The new rules and their enforcement also revealed political motives, not only on the part of the state but also on the part of a specific sector of the industry. Following a pattern consistent with developments elsewhere in the Portuguese economy, approximately 15 per cent of textile firms in 1946 (7 per cent in 1948 after changes in Industrial Regulation) conformed to the terms of the legislation. The circular favoured those grand textile manufacturers who dominated the textile associations, repeatedly complained about cutthroat domestic competition in their company reports, and lobbied for protected and colonial markets. It seems more than coincidental that the circular virtually gave this group *carte blanche* to expand or construct new factories if it so desired. Without actively eliminating smaller plants, the rules appeared to indicate that plants with under 10,000 spindles were undesirable. Similarly, preference for integrated units was given to those units which had a minimum of 300 mechanical looms.[8] Another circular in 1947 stated that requests for automatic looms from factories which had less than 100 looms would not be considered.[9] In effect, the government was condemning small firms to obscurity and favouring larger, more influential companies.

Further evidence indicates that on the one hand the government decided small plants were undesirable and on the other hand, favoured large firms. First, following the circular of 17 December 1946, many requests for new looms made by small factories were rejected. On 22 January 1947, the DGI announced 189 rejections along with a statement that all requests pending at the time of the circular which did not meet its conditions were denied.[10] Rejections continued through February, and by April, when it became clear that the government would not authorize new machinery for small factories, requests began to drop. Either small firms were not replacing machinery or they were not requesting authorization to do so.

[7] Bastos, *Comércio e indústria*, 163.
[8] *Bol. da DGI*, 'Despacho normativo' (17 Dec. 1946).
[9] *Bol. da DGI*, no. 487 (8 Jan. 1947), 'Indústria algodoeira'.
[10] *Bol. da DGI*, no. 489 (22 Jan. 1947), 331.

Large firms continued to receive authorization to replace or expand machinery as the Bulletin of the DGI (after 1949 DGSI) and company reports attest.[11] For example, the Fabrica de Fiação e Tecidos do Rio Vizela which had a total of 32,000 spindles and 1,213 looms in 1947 was given 180 days to install 8,000 new spindles.[12] The DGI repeatedly authorized requests by this firm for new machinery. The DGI also always approved the requests of Fábrica de Tecidos de Boa Vista de Guilherme Graham e Co. with 35,000 spindles and 1,000 mechanical looms and the Empresa da Fábrica de Fiação e Tecidos Oriental with 12,060 spindles and 1,213 looms. In addition, the Companhia Fiação e Tecidos de Alcobaça, with 14,000 spindles mentioned that it bought 18 looms in 1947, bought and modified machines in 1948, and continued to re-equip in 1949 and 1950.[13] The Cávado factory which had 12,000 spindles and 400 looms in 1947, replaced machinery in 1948, 1949, and 1950, according to company reports.[14]

The government extended its net over small firms one year after it passed the legislation on dimension. In 1947 changes in the laws defining cottage industries led to the inclusion of many small weaving establishments in the Industrial Regulation legislation governing the textile industry.[15] Whereas the statistics for 1947 revealed a total of 229 factories, the statistics for 1948 showed a total of 470 factories (including combined spinning and weaving mills and weaving only mills, see Table A13). Almost all of the increase was due to the incorporation of 234 weaving mills with an average of 7 looms per factory previously excluded from the survey, from the onus of requesting authorization for improvements and purchases, and from government surveillance of their production process. Even though these factories only added an extra 1,571 looms out of a total of 26,844 looms (leaving approximately 8,000 looms in the cottage industry outside the legislation), their abrupt incorporation into the Industrial Regulation system suggests that

[11] Many of the companies whose reports I read were followed randomly in the *Bol. da DGI* (after 1949, DGSI) from its creation in 1937 to 1965 and on no occasion do any of them appear to have been refused authorization. In a random sample of small factories between 1947 and 1949, almost all were rejected.

[12] See Carlos Bastos, *O algodão no comércio e na indústria portuguesa* (Oporto: GNIAR, 1947), Appendix for firm sizes quoted in this period.

[13] Alcobaça, *Relatório*, 1946–50.

[14] Cávado, *Relatório*, 1948–50.

[15] Rosas, 'Alguns aspectos', 5.

not only had the government dropped the populist element of its corporative stance and was no longer protecting, rhetorically or otherwise, small family-run concerns, but also that it was seeking their eventual elimination.

An examination of investment by large firms in the post-war period illustrates further that the government attempted to do for grand textile manufacturers what it was doing for the heads of financial-industrial conglomerates, though on a smaller scale and less consistently. Joint participation by the state, banks, and textile factories in the creation of the hydroelectric plant at Cávado and the establishment of textile factories in the colonies shows that the government was trying to effect vertical and horizontal integration in the industry. In 1944, following the passage of the National Electrification Law, the government began to finance the development of hydroelectric plants throughout the country.[16] Financed by both public and private capital, the plant at Cávado on the Douro promised to supply electricity more cheaply and reliably to the many textile plants in the surrounding area. Companies of limited liability were invited to invest alongside the state and banks. The company reports of Guimarães, Alcobaça, and Rio Ave testify that they and probably many more, took advantage of the opportunity to invest in the production of one of their major raw materials.

Furthermore, although the regime had refused to permit the construction of textile factories in the colonies since its inception, it overturned previous legislation in 1944 and issued Decree 33,924 allowing textile industries to locate in the empire.[17] Metropolitan textile manufacturers were invited to invest in the scheme but their companies had to be limited liability; the legislation therefore prevented small family businesses from investing or expanding into the colonies. Larger factories, the same ones who said they were replacing and/or increasing machinery, took advantage of the legislation and bought shares in one of the two companies actually established in the colonies.

The factory in Mozambique was located at Vila Pery and was owned by the Sociedade Algodoeira de Portugal. At completion it was to have 15,000 spindles and 500 automatic looms. The Angola plant was located at Luanda and had 24,000 spindles and 900

[16] Ribeiro *et al.*, 'Grande indústria', 950–1.

[17] D-L 33,924, *Bol. da DGI*, no. 368 (27 Sept. 1944).

automatic looms at completion. It was owned by the Companhia de Fomento Colonial which, in turn, was financed by the Banco Português do Atlantico (BPA). BPA also had an interest in Algodoeira Colonial Agrícola, a cotton concessionary company in Angola, and acquired a majority position in that enterprise in 1964. In 1950 the Sociedade Algodoeira de Portugal and the Companhia de Fomento Colonial merged to form the Sociedade Algodoeira de Fomento Colonial. By this time, the Luanda factory was already in operation and the Mozambique factory was near completion. The new company also took a major interest in coffee plantations in Angola, a hydroelectric plant at Revue in Mozambique, and part interest in a textile factory in Lisbon.[18]

The company received state funding through the National Development Fund (Fundo de Fomento Nacional) from 1950 and support from many of the leading metropolitan textile companies. Among others, the metropolitan companies which held shares were Empresa Fabril do Norte, Companhia de Fiação e Tecidos de Porto, Fabrica de Fiação e Tecidos do Rio Vizela, Fábrica de Fiação e Tecidos de Santo Tirso, Sampaio, Ferreira & Ca., and Companhia Fabril de Salgueiros.[19] In the early 1940s, *before the legislation was passed*, the Companhia Fabril do Cávado stated it had shares worth 200,000$00 in the Sociedade Algodoeira de Portugal, and in 1943 it invested another 1,000,000$00 in it. Similarly, after the legislation was passed the Companhia Fiação de Crestuma invested 464 contos and the Companhia Fiação e Tecidos de Alcobaça, 400 contos in the Sociedade Algodoeira de Portugal in 1944. The Companhia de Fiação e Tecidos de Guimarães reportedly gave 600 contos in 1945. In 1945 also, the Companhia Rio Ave reported it had 20 per cent of the capital in the Sociedade Algodoeira de Portugal. These companies and others contributed to colonial textile mills until the regime fell in 1974.[20]

It is not altogether surprising that metropolitan industries invested in the colonies: industries in England and France were also engaging in the same practice in their own colonies. What is interesting is that prior to the decree modernizing the industry, the government was already favouring the more powerful mill owners by allowing

[18] Sociedade Algodoeira de Fomento Colonial, *Relatório*, 1950.

[19] For a list of initial shareholders, see *A Indústria do Norte*, 93: 298–300 (Oct.–Dec. 1944), 14–18.

[20] See company reports for those years.

colonial textile companies to form in the face of a decree-law prohibiting textile industries in the colonies, then legitimizing what was in fact a *fait accompli* with the 1944 legislation, further assisting the process by allowing only limited liability companies, and finally, offering financial aid. Although the output of the colonial textile factories (once they were established) did not displace all metropolitan exports, these colonial mills did directly compete with exports from those companies which had not participated or could not invest in the colonial textile enterprise. Again, small firms were denied opportunities for expansion and investment by such policies. In every way, it seemed, the government in co-ordination with powerful manufacturers was seeking to undermine small industrialists.

THE CRISIS

By 1953, the year that Portugal launched its highly publicized First Development Plan, textile industrialists had begun to speak of a 'crisis' in the textile industry. As the decade progressed, almost everyone connected with the industry, from the President of the CRCAR to individual textile firms acknowledged that the industry was in severe trouble in spite of the 1946 legislation and government favouritism. Indeed, manufacturers disagreed little about the causes of the crisis. Variously described as a 'crisis of growth'[21] to a crisis of 'lack of business',[22] its manifestations were straightforward. Unstable or limited export opportunities resulted in the dumping of unsold textiles on an already over-supplied internal market, which led to lower prices and consequently, a decline in profits. As the Companhia Fiação e Tecidos de Alcobaça explained in its statement for 1953:

> In addition to the lack of business from which the crisis originated, there has arisen another aggravating circumstance: the foolishness of those who, desperate to dispose of their stocks, continually lower their prices, which upsets everything and everyone and renders the sacrifices we have already made counterproductive.[23]

[21] See statement by João Dias Rosas, in Part I of Armando Carneiro, *A crise da indústria têxtil algodoeira não e um mito* (Oporto: GEDES, 1958).

[22] Sociedade de Fiação e Tecelagem Rio Vizela, *Relatório da direcção, balanço e parecer do conselho fiscal*, 1958 (hereinafter Rio Vizela, *Relatório*).

[23] Alcobaça, *Relatório*, 1953, 6.

Over-production and underconsumption were working together to produce a vicious circle of oversupply, cutthroat competition, and price wars.

Although the characteristics were easy to identify, the way they worked as well as their origins were more complicated. Many of the factors which had resulted in over-production also contributed to underconsumption as well. It is perhaps the complexity of the problem which explains the government's eventual reluctance to resolve the difficulties of the industry or the industry's inability to combat them during the 1950s. Below we discuss the aspects and interaction of the various factors.

Over-production

When the President of the CRCAR called the crisis in the industry a 'crisis of growth' he was not far wrong. The industry had expanded enormously after the legislation of 1946. From 669,147 spindles and 24,519 looms in 1946 the industry had grown to 961,242 spindles and 31,914 looms by 1955 (Table A13). Moreover, integrated spinning and weaving plants had expanded from 65 in 1946 to 74 in 1951.[24] By 1951 they possessed 669,778 spindles out of a total of 835,940 spindles, approximately 80 per cent of the spindles in the industry; and 20,685 looms (compared to 17,703 looms in 1946), 70 per cent of the total number of looms. Combined spinning and weaving plants also possessed the majority of automatic looms, the most modern technology available, with 2,455 out of a total of 2,621 in the entire industry.[25] Thus not only had integrated units increased but also their possession of the majority of machinery indicated the persistence of an oligopolistic structure within the industry.

Part of this growth could be attributed to the effects of the 1946 legislation which encouraged expansion and modernization of larger textile mills. Anticipating the industry's expansion into foreign markets, the legislation sought to establish technical minimums on plant sizes to allow the industry to compete more effectively abroad. The availability of cheap colonial cotton, which it was thought would enable the industry to capture export markets also

[24] No breakdown of machinery by type of factory is available for 1955; 1951 figures have therefore been used. See Rosas, 'Alguns aspectos', Map 1.
[25] Ibid.

offered an incentive for factories to expand. According to the head of the CRCAR, this growth was not always well-planned:

the hope that it would be possible to depend on colonial cotton at relatively low prices and in sufficient quantities to operate without competition in national markets and in a privileged position in external markets provided an incentive to increase industrial capacity without careful planning by factories. This explains the abnormal development of the cotton textile industry in the post-war period. Presently, the working capacity of installed factories is much greater than the possibility of placement in markets. From that arises the difficulty of being able to work in economical conditions.[26]

The growth of more modern, well-equipped units had occurred, moreover, without the substantial decline of smaller units or the scrapping of antiquated machinery. Recalling the goals of the 1946 legislation, the President of the CRCAR reflected:

The objective of the circular consisted . . . in promoting the installation of well-equipped factories with efficient working conditions; compelling existing industry through healthy competition to improve its structure and elevate the technical and economic level of work and machinery; or even eliminating those factories which were unwilling or unable to accompany this development.[27]

In 1955—the year the notion of dimension was scrapped—out of a total of 104 factories in spinning, only 34 possessed the minimum of 10,000 spindles required by law; two-thirds of the factories did not meet the minimum requirements. In weaving, although 56 mills had disappeared between 1948 and 1955, 83 per cent of weaving firms still had less than 100 looms and the total number of looms in those mills had risen slightly. The legislation had been unable to eliminate those at whom it was directed even though growth and modernization had occurred and some factories had closed. Because growth had taken place in an already overcrowded market, textile manufactures increased, prices fell, and a crisis ensued. Even in 1954, a year in which exports abroad reached their highest level ever, the stock of unsold textiles still equalled 5,852 tons.[28]

[26] Ibid. 12.

[27] Ibid. 5.

[28] C. Corte-Real, 'Indústria têxtil algodoeira', in *Discursos, conclusões e estudos sobre a indústria portuguesa*, II Congresso da Indústria Portuguesa (Lisbon: AIP, 1957), 6.

How were small firms able to survive in the face of legislation designed to secure, or at least, hasten, their demise and in a market which was oversupplied and in which they had to compete against larger, technically advanced firms? No single explanation suffices, the reasons can be attributed to the following factors, all of which worked together: cotton supply, its price, and its method of distribution; government inability or unwillingness to enforce the legislation or secure new markets; low wages and over-employment; and the perseverance of small factories, whose capital costs had long since been amortized.

Paradoxically, colonial cotton shortages and expanding needs of larger mills during the 1950s probably allowed smaller mills to survive. Even in years of known shortages, all colonial cotton was distributed first through the quota system: each importer was entitled to so much cotton based on past purchases. Importers then allocated cotton to factories (see Ch. 6). Sometimes, all the colonial cotton would be distributed and factories would not have received their full amount of cotton, either because there had been a drop in colonial cotton production or because individual quotas assigned to their importers had not been revised in line with the expansion of the industry. Those left with insufficient supplies then had to purchase higher priced cotton on the international market. Any advantage that larger firms enjoyed by being more efficient than older, smaller firms was offset by the need to buy more expensive, foreign cotton. One factory explained:

> As predicted in our last report, in the course of the year which just ended, our business like others only received a contingent of colonial cotton corresponding to little more than half of the quantity necessary to work to capacity. In order to avoid a reduction of work hours or the paralysis of a spinning shift which would have had deplorable consequences for our workers, we had to resort to the purchase of foreign cotton at a much higher price.[29]

In 1952 another year in which the industry had to resort to foreign imports due to shortages and/or expansion, the Companhia de Fiação e Tecidos de Alcobaça also referred to the increase in price of their textiles and a subsequent drop in sales as a result.[30] Throughout the 1950s, companies repeatedly named the lack of

[29] Fábrica de Fiação e Tecidos de Santo Tirso, Lda., *Relatório*, 1951, no page.
[30] Alcobaça, *Relatório*, 8.

colonial cotton as one of the major contributors to the crisis. The Sociedade de Fiação e Tecelagem Rio Vizela said in 1956:

> The lack of overseas (colonial) cotton recorded during the year necessitated the acquisition of higher priced foreign cotton, aggravating the difficult life of the textile industry. For several years the industry has struggled with rampant competition which has had disastrous effects upon various firms, who are unable to endure the resulting crisis.[31]

Despite the over-production already evident in the industry, Rio Vizela viewed modernization as a possible solution and announced it was going to buy 8,000 new spindles to replace its old ones. It also expressed the hope that taxes (which had steadily increased since the close of the war) would be lowered to alleviate the effects of the crisis.

In 1957 the crisis continued. The company reported:

> The difficulties referred to in previous statements, deriving from the lack of overseas cotton and uncontrolled competition in the sale of products, continued seriously to preoccupy the directors of the textile industry during the year. They were compelled to adopt technical and economic measures to overcome this long crisis yet without satisfactory results.[32]

The Fábrica de Fiação e Tecidos de Braga reiterated these complaints and fears when it mentioned difficulty in product placement, rampant competition, and price drops. Additionally, the Fábrica de Fiação e Tecidos do Jacinto lamented the fact that prices in the market continued to drop even while production costs were rising:

> In addition to the increase in the cost of colonial cotton, we were surprised at its scarcity, which forced us to acquire more expensive foreign cotton in order not to restrict our working days to two a week. On the other hand, a situation which is already sufficiently aggravated has tended to worsen as a result of price increases of nearly all the materials used in our manufacture.[33]

In 1958 it referred to the price cutting practised by others:

[31] Rio Vizela, *Relatório*, 1956, 8. In accordance with the 1951 constitutional change, many factories refer to colonial cotton as overseas cotton. I am preserving the distinction in my translation.

[32] Ibid. 1957, 7.

[33] Fábrica de Fiação e Tecidos do Jacinto, *Relatório*, 1956, no page (hereinafter Jacinto, *Relatório*).

Within the industry the notion prevails of selling cheaper than the competition. While this state of ideas is maintained, the already long crisis that we have been going through will continue, and if our government does not resolve to study it and find a solution to it, it is our opinion that many poor days will arrive.[34]

Other large firms studied in this period stated that they did not receive remunerative prices for their products and referred to uncontrolled or unhealthy competition. Principally, these larger firms found themselves competing in the domestic market against smaller, antiquated firms which had bought cheaper colonial cotton through the quota system. Since smaller firms were able to meet most of their needs through the purchase of cheaper colonial cotton they had an advantage over those firms which had to rely on foreign cotton imports. Small firms were able to sell their products more cheaply in the domestic and colonial markets which may have contributed to their survival. The vast difference between the price of colonial cotton and the international price, the continued intermittent reliance on foreign cotton to compensate for colonial shortages, and the ability of the small firms to use the quota system to advantage initiated much talk of setting an average price for colonial and international cotton but this never occurred.[35]

Since colonial cotton still comprised the majority of total raw material supplies during the 1950s, other factors must have accounted for the ability of small mills to survive in the domestic market. These factors were: the cost of labour, tax avoidance, illegal sales, and low fixed capital costs. While wages were appallingly low in the textile industry as a whole, small weaving firms in particular benefited from the use of family labour to keep wages at a minimum or to delay the payment of wages when prices were low. One observer reckoned that 30 per cent of mechanical looms in the industry were employed in units where workers were paid terribly low salaries. Added to these regulated firms were cottage mills employing approximately 8,000 looms not covered by the Industrial Regulation legislation, which worked on commission and competed with the bigger factories. Located in rural areas, they worked in unhygienic conditions at low pay (6$00 to 9$00 escudos a day) and long hours (10 to 12). Women and children working on

[34] Ibid.
[35] Rosas, 'Alguns aspectos', 27.

old manual machines comprised the majority of the labour force. Moreover, many of these small factories, both regulated and unregulated, allegedly bought on credit, avoided taxes, and sold their goods illegally at fairs or unofficial markets.[36]

It was estimated, furthermore, that 300 mills with approximately 1,300 looms, which had been officially classified as artisanal factories, were in reality not engaged in specialized or artisanal work, but were competing with the rest of the textile industry.[37] While regional and artisanal activity was encouraged, the rest of the industry deplored this kind of competition. Finally, unable or afraid to purchase new machinery because of the 1946 circular and/or their precarious financial conditions, many small factories were working with machinery which had long been amortized so fixed capital costs were low. Even if small firms were struggling, these factors, combined with the colonial cotton quota system, enabled them to compete in an already overcrowded market.

Most evidence suggests that the government was indeed interested in the technical regulation of the industry as it was with regard to every other Portuguese industry during this period. General and specific legislation, Salazar's speeches, the priorities of the First Development Plan—all depicted a country which was trying to improve technically, to develop industrially, and to increase its exports. But as we witnessed in the previous chapter, the regime enforced its policies inconsistently, or the policies produced contradictory outcomes and had unforeseen consequences. In textiles, the regime was unwilling directly to enforce the legislation it had so vigorously promoted. It made few moves to shut clandestine factories or those which had failed to meet the minimum requirements. Moreover, the government did not try to hold falling domestic prices by setting minimum (as opposed to maximum which it did set) price levels nor did it offer subsidies to firms selling textiles on the domestic market made with foreign cotton (even though firms had been forced to buy the more expensive cotton through the operation of the quota system).

One does not have to search far to find the reasons for government disinterest and reluctance. It is pretty clear that textiles were no longer a matter of top priority for the government; the industry

[36] Armando Carneiro, *Antelóquio de um inquérito económico-social feito à indústria têxtil algodoeira* (Vila Nova de Famalicão: Tip. Minerva, 1959), 36–7.

[37] Rosas, 'Alguns aspectos', 36–8.

was losing importance during the 1950s as non-traditional industries began to make their mark in Portugal. Secondly, closing those factories which did not conform to the 1946 circular would have been politically difficult. The regionally based, textile industry employed 70,000 workers. In 1955 the district of Braga had 40 per cent of the total labour force in the industry in 227 factories, the majority of them small; another 50 per cent worked in the contiguous district of Oporto, in 189 factories both large and small.[38] According to one estimate, a 20 per cent drop in production would have left 12,000 to 15,000 workers unemployed.[39] Given such a concentration of mills and labour, shutting many small factories in the Minho would have created a degree of social disruption which the government could not have wanted to risk.

Thirdly, given the regime's rhetoric on the family and small business, it would have been difficult for the government to move openly against smaller, older factories. This would have provided political ammunition to the growing opposition movement in the country during this period. The support for Norton de Matos in the 1949 presidential election may have drawn the government's attention to the problem of social protest: the number of rejections on machinery applications to the DGSI (formerly DGI) dropped after this date.[40] Finally, the failure to tackle the cotton problem lay in the government's persistent preference for stability. If it subsidized foreign cotton imports or raised the price of colonial cotton, the logic of colonial protection disappeared and the government risked unforeseen consequences for the economy as a whole. Unfortunately, this narrow attachment to price stability had its own drawbacks as events after 1958 confirmed.

Underconsumption

Since the number of mills had not decreased and textile production had rapidly expanded, markets became crucial. Although both internal consumption and external markets had improved, they had not kept pace with textile production and growth, nor was demand stable. Average textile consumption in Portugal was only 3

[38] Francisco Pereira de Moura *et al.*, 'Estudos sobre a indústria portuguesa', in *Discursos, conclusões e estudos sobre a indústria portuguesa*, II Congresso da Indústria (Lisbon, AIP, 1957), 98.

[39] Rodrigo Carvalho, quoted by Bastos, *Indústria e arte têxtil*, Appendix, x.

[40] Brito, *A industrialização*, tables 5.10–5.12.

kilograms per capita compared with 3.3 kilos per capita in Italy, 4.7 kilos in Denmark, and 6.6 kilos in Switzerland and Great Britain; this was attributable to the low salaries paid to Portuguese workers, particularly those in agriculture.[41] For the low wage situation, grand industrialists in the textile industry are partly to blame for they blocked government attempts to raise wages just after the war.[42] Per capita consumption in colonial markets was even lower and additionally, two textile factories had begun operation in Angola and Mozambique so the market there remained stagnant throughout the 1950s. Average annual sales in the domestic market between 1950 and 1955 were 22,060 tons and in the colonial market, 9,899 tons.[43] These markets were unable to absorb all that the industry produced.

Exports abroad thus became imperative, but in the post-war years these tended to be unreliable and to oscillate from year to year. One of the reasons for this was that textile industries were improving in other countries causing a drop in demand for textile imports from elsewhere; but reasons also lay in government disinterest as well as the quality and price of Portuguese textiles. The ostensible purpose of the 1946 circular had been to improve the industry's chances of competing abroad, yet the government made little effort to secure trade agreements for textiles.[44] Its attention was focused on the demands of the conglomerates; textiles as a traditional industry was not a top priority. Equally, despite growth, the industry itself was ill-prepared to compete abroad as the discussion below reveals.

First, the ever-present difficulties with colonial cotton affected the industry's chances in the more demanding export markets. Even in years when the overall quantity of colonial cotton was sufficient, procurement was unstable and disorganized.[45] Weavers

[41] Corte-Real, 'Indústria têxtil', 6.

[42] Fátima Patriarca, 'O triângulo corporativo: Acta e encenação de um despacho salarial (1946–1947)', *Análise Social*, 23: 99 (1987), 905–44, for an account of the battle over wages among the state, large textile manufacturers, and workers.

[43] Moura *et al.*, 'Estudos sobre a indústria portuguesa', Appendix, table VI.

[44] Maria Helena Garcia da Fonseca, 'Intervencionismo do Estado no sector têxtil algodoeiro', Estudos no. 6, Instituto Nacional de Investigação Industrial, 1962, 131. D-L 40,276 (11 Apr. 1955) exempted duties on imports of Brazilian cotton (which eased the situation for larger firms) and also suspended export duties on goods made with Brazilian cotton. Fonseca has no evidence on trade agreements with other countries until the 1961 EFTA agreement.

[45] Rosas, 'Alguns aspectos', 13.

never knew from whom they were buying or in what quantity. Often, the worst quality seemed to be in the greatest supply. Erratic supply also meant the quality of cotton was often improperly matched to the type of textiles produced. Fine quality cotton or thread was used for articles of medium or coarse quality and vice-versa, with the result that orders from abroad were cancelled or not renewed.[46] Reclassification and the proper matching of raw material to product was necessary, but difficult to achieve under the primitive quota system.

Secondly, the quota system seemed highly arbitrary. Whereas some factories had quotas for three shifts, others had quotas for only two or even one shift.[47] For this very reason, the quota system also offered a disincentive to expand precisely because larger factories were unsure whether their quota would be increased or by how much. Also, large factories did not know what expenses would be involved should they need to depend on the purchase of more expensive foreign cotton.

Thirdly, despite low wages the Portuguese textile industry still had high costs of production thus forcing up the price of their textiles. Even though many firms spoke of their modernization efforts, low productivity and factories working in unsuitable conditions predominated. The workforce was uneducated and did not possess the skills to improve productivity.

Fourthly, compared with other countries, the density of workers in both spinning and weaving was high. Whereas in Great Britain and France the ratio of spindles per worker were 150 and 179 respectively; in Portugal, it was 67. For looms, the ratio of looms to workers was over 2 to 1 in Great Britain and France and 1.5 to 1 in Italy; in Portugal, it was 1 to 1.[48] Unproductive use of labour offset some of the advantages of low wages. Again, this issue related to general government policy which preferred to see over-employment in the textile industry cancel out the effects of low wages, rather than face the problems of restructuring, unemployment, and unionization.[49]

Fifthly, old and unproductive machinery still existed. The average age of spindles including those in integrated firms was 25 years old

[46] Ibid. 16.
[47] Carneiro, 'A crise', Part II, 70.
[48] Moura *et al.*, 'Estudos sobre a indústria portuguesa', 105.
[49] Rosas, 'Alguns aspectos', 15.

in 1955 and 30 per cent of spindles in the entire industry were over 30 years of age. As for looms, it was estimated that approximately 22,215 looms out of a total of 31,914 were over 20 years old.[50] Moreover, the ratio of spindles to looms, at 20 to 1, was still low compared with other countries, and the yield per machine was low. Taking 300 days of work a year as a normal working year, the average use of machines was 51 per cent for spindles and 40 per cent for looms.[51]

Finally, connected to the problem of low yields was the way production was organized. In order to satisfy a diversity of orders, factories would sometimes work certain spindles three months out of the year solely in order to produce different qualities of thread and cloth ranging from coarse to fine quality. While some variation was an accepted component of an industry such as textiles, a factory committed to varied production was inefficient because many machines saw little use for long periods.[52] The episodic working of machinery gave an advantage to those small firms with ancient plants, with low fixed capital costs, and amortized machinery as opposed to those who had recently purchased modern machinery.

Because of these factors those companies geared towards exports to foreign markets (about eighty-six in 1958) found that they were not competitive abroad. Low colonial cotton prices in the years when supply was sufficient probably helped to offset some of the difficulties and disadvantages, but did not eradicate them.

No wonder then that large firms which had modernized and expanded also spoke of a crisis in these years. Not only had they been undermined in the domestic market, but a combination of factors had reduced the export market. Yet simultaneously, many manufacturers felt that replacement of old by new machinery was the only way to overcome the crisis.[53] With such drops in prices and extensive competition—presumably what the government had sought to avoid through price controls and Industrial Regulation since the Depression—such expansion or modernization may have been impossible for many.

[50] Moura *et al.*, 'Estudos sobre a indústria portuguesa', 97.

[51] Carneiro, 'A crise', 68; and Corte-Real, 'Indústria têxtil', 7. According to Corte-Real, Portuguese workers worked even less than this average.

[52] Rosas, 'Alguns aspectos', 16.

[53] See e.g. the statements for the late 1950s by the Braga, Cuca, Rio Ave, and Tomar mills.

Who bore the responsibility for high production costs, disequilibrium, and lack of markets—in short—the crisis in the industry? Neither the industry nor the government can escape blame. Low colonial cotton prices and ready markets during World War II brought substantial profits to the industry as we documented. In the post-war period, many manufacturers took advantage of their good fortune to modernize and expand. But others, as one manufacturer noted, spent their profits on 'luxurious purchases which offended public opinion'[54] and this had led to lack of finance for investment in the 1950s. It was equally true that with low wages and a protected, lucrative market in the colonies there was insufficient incentive to be zealous about modernizing.

Yet the government bears a share of the responsibility as well. It did not enforce substantial wage raises which would have compelled the industry to modernize and would have led to increased domestic consumption. It made little effort to negotiate trade agreements which would have secured export markets for the industry. Worst of all, it passed legislation which was unrealistic and inappropriate for the Portuguese textile industry. Although the 1946 legislation was neither systematically nor consistently enforced, it stifled modernization for smaller firms. According to one manufacturer:

> the idea of dimension—badly defined and badly directed—appeared to industrialists sometimes as a suspect means of promoting monopoly concentration and to others simply as an inducement to create large textile units capable of reducing fixed expenses by higher production and lower manufacturing costs, thereby affording better opportunities of competing in the international market.[55]

Thwarted by the 1946 circular, at least 56 weaving mills collapsed; others were prevented from buying modern machinery even assuming they were so inclined and could afford it. Part of the difficulty was that the 1946 circular imposed static, administratively convenient solutions to rapidly evolving situations. Textile markets, colonial cotton supplies, the world price of cotton, and textile machinery were unstable; the notion of dimension was subjective. The strait-jacket of regulations created as many prob-

[54] Lino Lima, *A crise da indústria têxtil algodoeira* (Vila Nova de Famalicão: Centro Gráfico, 1967), 13.

[55] Ibid. 14.

lems as it resolved. Equally, the scheme had been introduced when Portugal was meeting 95 per cent of its needs through the supply of colonial cotton and had not considered the effects of industrial expansion and possible colonial cotton shortages, nor the ability of small firms to play the quota system to advantage against bigger firms.

The government withdrew the legislation in 1955. Under the terms of a circular issued on 17 December 1955 the government nominated a commission to study the present structure of the industry, investigate the economic and technical conditions under which it worked, and offer suggestions for the future improvement of the industry.[56] The circular also established new guidelines for growth in the industry. According to the 1955 circular, authorizations on the expansion or replacement of machinery would be granted according to the following criteria. Decisions on present and future requests would consider the 'stable expansion and modernization of factories (particularly small and medium-sized mills), as long as they were technically and structurally sound and such modernization or expansion did not have negative social and economic consequences'.[57] The circular also sought to improve working conditions in the industry in order to bring it up to the standard of modern textile industries in other countries. While the legislation probably came as a relief to smaller industries, its terms were sufficiently vague, suspiciously close to the goals of the 1946 legislation, and allowed the government ample opportunity to intervene if it desired. Three years after the passage of the legislation, statistics show that the number of mills declined overall by 17; those mills with less than 30 looms declined by 19 between 1955 and 1958. The majority of expansion took place in factories which had over 10,000 spindles and 300 looms (see Tables A13–A15). The shadow of the 1946 legislation persisted.

CONCLUSION

From formulation to implementation, post-war policies for the textile industry suffered from a succession of errors. After the war,

[56] *Bol. da DGSI*, 8: 374 (29 Feb. 1956), Despacho (17 Dec. 1955).
[57] Ibid.

the regime not only wished to favour a small group of powerful textile manufacturers but did so in order to expand into international markets and to concentrate textile manufacturing as it had done for other branches of Portuguese industry. However, in a country dominated by small firms, the notion of dimension adopted in the 1946 legislation was inappropriate and served only statically technical and blatantly political aims.

In practice, no part of the legislation was faithfully executed. Government policy overall was inconsistent and contradictory. When small industry was able to survive in the domestic and colonial markets by taking advantage of the quota system and relying on low wages and the black market, the regime was unwilling to use the powers it had given itself to close small plants for fear of the repercussions in Minho. That left a marginal sector alongside a more highly capitalized one. The former persistently undercut domestic prices, depriving the latter of a reasonable domestic base to launch its exporting activities.

The government pledged to assist larger firms but did so only half-heartedly. Higher quality cotton was not obtained and periodic shortages of colonial cotton were not resolved. Domestic consumption did not improve neither were export markets secured. In 1958 companies continued to speak of cutthroat competition within and the industry was in a precarious state. More than a decade after the industry was poised to embark on a period of modernization within and expansion abroad, the Directors of one textile company spoke for many when they declared, 'we fear that if methods are not officially taken to defend this industry, one of the largest in the country, unprofitable years will continue'.[58]

[58] Companhia da Fábrica de Fiação de Tomar, *Balanco, relatório da direcção e parecer do conselho fiscal*, 1958, no page (hereinafter Tomar, *Relatório*).

9

Colonial Cotton in Transition

Active instruments of colonial policy from 1946 to 1958 were the interconnected, mutually dependent schemes of development, mechanization, industrialization, and European settlement. They were closely related to events occurring in the metropole at this time: an increasingly technocratic orientation and the development and modernization of industry. The colonies were to be integrated into this scheme, not only through the absorption of excess capital, but also by the settlement of emigrants from Portugal, often illiterate and unskilled. The policy outlines were visible as early as 1943 but only really implemented after the war.

As the metropole modernized and mechanized so the colonies were expected to receive people who had been displaced by modern technology or industrial concentration. Moreover, to face greater post-war competition, the government exhorted the metropole and the colonies to produce more at better prices, in order to appease settlers, deter possible foreign imports, and compete in export markets. Mechanization would make this a possibility. Finally, the regime saw the colonies as areas of investment for favoured industries which had acquired capital during the war and were now looking to expand. The period was one of increasing ties between the colonies and the metropole. Later, the strength of those ties would be severely tested.

Just as development, mechanization, and labour were discussed for the colonies as a whole so also were they discussed within the cotton regime. Since the metropolitan textile industry was rapidly expanding in the early 1950s, boosting productivity and quality to meet increased demand became a major objective of the regime and cotton concessionary companies. Yet, while the state relied mainly on coercion to fulfil its policy objectives until 1946, in the post-war period it began to downplay coercion in favour of material and moral incentives to producers and companies and to make improvements in cultivation, transport, and storage. Legislation in the period attests to the multiple strategies employed—from social engineering to soil studies—to improve the cotton regime in Angola and Mozambique, but the results show that the new policies were not entirely effective. Additionally, these efforts took place during a decade of falling international prices, a situation which by 1958 would challenge the state's allegiance to its own policy of price-fixing and threaten to undermine the government itself.

GENERAL DEVELOPMENTS

The 1951 constitutional reform legally and politically integrated Mozambique and Angola into the Portuguese nation-state as Chapter 7 outlined. Concomitantly, administrative reorganization took place within each colony. Through these changes the Portuguese government sought to deflect settler and international criticisms of Portugal as a colonial power, while fulfilling metropolitan economic objectives. As we have seen, Portuguese capitalists made large profits out of the war which they were then anxious to invest. The Salazar Government responded by passing legislation which offered credit and state assistance for several development projects in the metropole including hydroelectricity and steel. Also, the government formulated plans for colonial industrialization to give Portuguese capital the incentive to invest in the colonies. Government promised to aid this process by providing infrastructure, energy, and communications through the development scheme. Such favours were not disinterested, the Salazar regime needed to build political support at home and in the colonies. It wanted to conserve foreign exchange through import substitution and a reduced reliance on foreign imports, and it sought to fortify the nation against possible foreign competition. As the Minister of the Colonies declared in 1948, 'The colonial policy of attributing to the overseas territories the simple role of supplier of raw materials to the metropole has had its day.' Moreover, the colonies could no longer be treated as a 'market of refuge' where Portugal could dump its overpriced, poorly made goods. For Portugal to stay in step with other colonial powers it needed to industrialize its empire.[1]

The new approach to the colonies began as soon as the war

[1] *BGC* 24: 278–9 (Aug.–Sept. 1948), 100.

ended. In Angola, the 1945 Development Plan allocated 300,000 contos to the development of hydroelectric power, agriculture, forestry, mining, and communications. The hydroelectric scheme also received half a million dollars of Marshall Plan money. In Mozambique, the Limpopo Valley Irrigation scheme intended to make available for European cultivation a swathe of territory in the south. Furthermore, in 1947, the government granted loans both to Mozambique and to Angola to cover improvements or extensions of infrastructure, especially ports and railroads, communications, and the supply of energy.[2]

The regime also included both colonies in the First Development Plan of 1953. In Angola and Mozambique, the plan endeavoured to develop electrical energy, communications, ports, railways, and oil and mineral exploration. An integral part of the plan was the mechanization of agriculture. As with the development strategy, the emphasis on mechanization was not only an admission that the Portuguese had money to invest but also that times had changed. It was necessary to lower costs of production and be more competitive. For years African labourers, using the simplest of cultivating tools—their hands—and employing agricultural methods imposed on them by Europeans had been depended upon to harvest the agricultural produce of Angola and Mozambique. When post-war competition began to loom, several spokesmen for the colonies worried about the dangers a lack of mechanization would bring. As a respected cotton engineer argued:

> So long as the cultivation of cotton in the Portuguese colonies continues to be accomplished manually, we can be certain that it will be impossible to increase production and decrease its cost appreciably, to compete alongside foreign cotton in the markets, or to elevate the standard of living of the producers. Those who continue to advocate this system of work say that the labour of the 'natives' is cheap and abundant. But anyone who wishes to verify the number of days which are necessary to cultivate a hectare and the ridiculous harvest gathered with such a faulty and 'primitive' system of labour, must reach the conclusion that a kilo of Portuguese cotton costs more than the same weight harvested in North America, Egypt, or Brazil.[3]

Connected to the organization of the cotton system itself which exploited Africans and led to resistance, the inefficient utilization of

[2] Ibid.

[3] Ribeiro, *Da colheita*, 100. Here and elsewhere, 'natives' and 'primitive' are literal translations from the Portuguese.

manual labour was affecting the quantity and quality of cotton produced. Cotton harvesting by machine was less costly, wasteful, and time-consuming.

This view was voiced for other products as well. After the war, producing and producing well became high priorities fuelled by fears of international criticism and competition. One writer put the case for mechanization succinctly:

> Without a minimum of mechanization, with only the manual labour of the 'native' or oxen for ploughing, it is not possible to go very much farther as far as quantity, quality, and prices than we have already done.[4]

However, only in the mid- to late 1950s did the real drive to mechanize begin.

The metropolitan government's scheme for colonization developed alongside growth and mechanization plans. The colonies would be used to receive Portugal's excess population, mostly rural, unemployed and illiterate. In 1947 the government revealed the seriousness of its intent when 1,000 emigrants left for Angola, causing one writer for the *Boletim Geral das Colónias* to remark that 'such a high number of settlers had not previously embarked [to the colonies]'.[5] In order to relieve unemployment in Portugal, strengthen Portugal's claim over the colonies, and make its control more effective, the regime had returned to the old policy of exporting 'excess' population to the colonies and resettling it in distant places, yet with little infrastructure and few social services. The settlers were expected to engage in agricultural activities and cotton production was among those pursuits encouraged. In 1949, 32 European producers were reportedly engaged in cotton production; by 1950, the number had risen to 45. The arrival of Europeans would have an interesting effect on cotton production, particularly in Angola in the 1960s and 1970s.

The intention behind the colonization drive was that these settlers would contribute to the advancement of Angola and Mozambique and bind the colonies to the metropole. Settler development schemes were implemented in the Limpopo Valley and Zambezia region in Mozambique; at Cela in the Central Highlands and Matala in the Cunene River Valley of Angola. Portuguese farmers

[4] A. C. Valdez dos Santos, *Perspectivas económicas de Angola* (Lisbon: Agência Geral das Colónias, 1949), 229.

[5] *BGC* 23: 264 (June 1947), 47.

were given supplies, land, housing, subsidies, and technical advice to start farming projects, either on their own or as a part of these schemes.[6]

Africans were also a part of the government settlement scheme. As we saw in relation to cotton production, many Africans attempted to avoid compulsory cultivation through migration to other parts of the Portuguese colonies or to other countries such as South Africa. During the 1950s, Africans also began to migrate to the large towns and cities which had begun to arise in Mozambique and Angola. The motive for the establishment of *concentrações* (model villages) and co-operatives in this period was to eliminate migration, create and maintain a stable African labour force. As others were doing at this time, Portugal was constantly searching for the best method of settling labour and the easiest and most profitable method of encouraging Africans to produce and ultimately, to consume. One was through grouping Africans on designated areas of land and literally supervising their production—the *concentração*.

The other method was by allowing Africans to form co-operatives to reduce costs and increase their trading power. Co-operatives for Africans were created in various parts of Angola and Mozambique in the early 1950s. In the *circunscrição* of Chibuto in Mozambique, 16,232 Africans were formed into co-operatives and in the Limpopo Valley, another 1,200 Africans were reportedly settled on 5 to 6 hectares of land each; 5,000 farmers in Gaza province were in co-operatives and three associations of Africans in the 'pre-co-operative' phase were in Manica province by 1952. These associations became fully functioning co-operatives by 1957.[7] It is not known if any of these Mozambican co-operatives included cotton producers. Since the majority of these co-operatives were in the southern part of Mozambique, it suggests that the government was attempting to encourage male Africans to stay in Mozambique rather than migrate to South Africa or, at least, to establish settled households in which they would invest their remittances from migrant labour. Though the Mozambican colonial government profited greatly from the migrant labour scheme, it left the southern area of Mozambique depopulated. Estimates indicate that 100,000 Mozambicans migrated annually to South Africa in the 1950s and

[6] Gerald J. Bender, *Angola under the Portuguese: The Myth and the Reality* (Berkeley and Los Angeles: Univ. of Calif. Press, 1978), 104–5.

[7] *BGU* 28: 329 (Nov. 1952), 128–31; and *BGU*, 33: 379 (Jan. 1957).

1960s. The Mozambican authorities, tired of complaints by companies of constant labour shortages and dogged by the threat of famine in the south, were anxious to increase labour and food production.[8]

RESULTS OF COLONIAL DEVELOPMENT PLANS

Both colonies experienced growth in agricultural production during the period. In the 1950s, Mozambique posted average annual growth rates of 7 per cent and Angola, growth rates of 5 per cent.[9] Angola also showed substantial mineral production in this period. Its greatest success was oil, discovered in 1955. Coffee production, one of Angola's largest agricultural exports, grew enormously during the early 1950s when a commodity shortage brought on by the Korean War sent prices soaring. Whereas in 1948 the price of coffee was 8$61 per kilo; in 1954 it was 29$39 per kilo.[10] Through coffee exports Angola was able to build up a favourable balance of trade with the United States, a matter of great satisfaction, no doubt, for the Portuguese government.[11] The increase in price led to expanded coffee production and subsequently of exports—from 53,400 tons in 1948 to 118,000 tons by 1961.[12] The negative effect of the boom, however, was that it led to the displacement and consequent social disruption of Africans from northern parts of Angola and the confiscation of their farms by Portuguese settlers and businesses eager to take advantage of the boom. In 1955 alone, 5,000 Africans in Uíge and 10,000 Africans in Dembos were moved off their lands as coffee cultivation expanded.[13] Where quick profits could be made, the Portuguese apparently worried little about settling Africans.

Other crops which enjoyed growth in Angola after World War II were sisal, corn, cotton, and cattle. Mozambique, though it did not

[8] First, *Black Gold*, 32–3; and Pitcher, 'The Agrarian Question', 38–9.

[9] Frank Brandenburg, 'Development, Finance, and Trade', in D. M. Abshire and Michael A. Samuels (eds.), *Portuguese Africa: A Handbook* (New York: Praeger, 1969), 227.

[10] Henderson, *Angola*, 135.

[11] Great Britain, Board of Overseas Trade, *Report on the Economic Conditions in Angola*, 1953 (London: HMSO, 1954), 27.

[12] Ibid. 135.

[13] Bender, *Angola under the Portuguese*, 166 n. 21; and Henderson, *Angola*, 135.

develop on the scale of Angola, also experienced an increase in output. Its major crops were sugar, copra, cotton, and cashew-nut as well as invisible earnings from the use of transport facilities and the export of migrant labour to South Africa (whose deferred payments were made in gold after a 1940 agreement).[14]

With regard to trade, the percentage of imports from Portugal into Angola remained at around 48 per cent up to 1960, while 30 per cent of imports into Mozambique were Portuguese during the 1950s. Over the same period, both colonies consistently imported more from Portugal than they exported to it (except 1953–4 for Mozambique). More importantly, Angola had a healthy trade balance with third countries. Mozambique consistently had a negative balance of trade with third countries in addition to its negative balance with Portugal but the 'invisible' income it received from the export of labour to South Africa and the use of its transport facilities by other countries compensated for its poor trade balances, contributing valuable foreign exchange to the empire's finances.[15] According to Xavier Pintado:

> it is the surpluses of the balance of payments of the overseas provinces which as a rule have more than re-established equilibrium in the payments of the escudo area and have made for the regular increase of the country's gold and foreign exchange holdings.[16]

Government-sponsored plans and investment by private companies also achieved some success, though neither industrialization nor mechanization made appreciable advances as a result. Although it neglected health and educational services, the 1953 Development Plan in the two colonies did generate expansion of infrastructure and hydroelectric schemes. According to M. D. D. Newitt, 'the amounts invested in the colonies were broadly comparable with those invested by other colonial powers.'[17] The *Boletim Geral do Ultramar* reported in 1956 that investment in Mozambique by ten companies had grown from 214,803 contos in 1939 to 751,837

[14] First, *Black Gold*, 25.

[15] Brandenburg, 'Development, Finance', in Abshire and Samuels (eds.), *Portuguese Africa*, 242 and 248; and on the contribution of gold to Portugal's balance of payments, see also Joana Pereira Leite, 'La Reproduction du réseau impérial portugais: Quelques précisions sur la formation du circuit d'or Mozambique/Portugal, 1959–1973', *Estudos de Economia*, 10: 3 (Apr.–June 1990), 365–401.

[16] Pintado, *Structure and Growth*, 234.

[17] Newitt, *Portugal in Africa*, 221.

contos in 1954—an increase of 251 per cent.[18] Almost all of the financial-industrial conglomerates had holdings in the colonies in agriculture, minerals, insurance, industry, banking, and the import–export business. With financing from the conglomerates, refineries were constructed in Angola and Mozambique following the discovery of oil in Angola. Iron-ore production also began in Angola from 1956.[19] These investments translated into increases in electrical output, cement and industrial production, and transport facilities during the 1950s.

Efforts to settle Europeans had more mixed results. Whereas in 1940, there were 27,400 whites in Mozambique and 44,000 in Angola, by 1950, 48,200 whites resided in Mozambique and 78,826 in Angola.[20] However, for the most part the settlement schemes were unsuccessful and cost the government substantial sums of money. In Angola, 1,700 families were supposed to have been settled at Cela and 1,000 at Matala during the 1950s. By 1960 only 300 families had been settled in each scheme at an approximate cost of $100,000 dollars per family in Cela and $25,000 dollars per family in Matala.[21]

Settlement schemes for Africans fared even worse. Europeans and companies resented the loss of African labour from their farms and enterprises. On the whole, most of the schemes were also badly co-ordinated, poorly managed, and made no provision for the social well-being of the people involved. According to Gerald Bender, the situation in Angola was such that 'instead of receiving new houses, furniture, animals and irrigated land, "settlement" for Africans was nearly synonymous with the loss of their best lands'.[22]

POST-WAR CHANGES IN THE COTTON SCHEME

The cotton scheme was included in the post-war changes sweeping the colonies. With the domestic textile industry in a dilapidated state and only just beginning a programme of modernization,

[18] *BGU* 32: 370 (Apr. 1956), 114.

[19] Abshire, 'Minerals, Manufacturing, Power and Communications', in Abshire and Samuels (eds.), *Portuguese Africa*, 298.

[20] Newitt, *Portugal in Africa*, 164.

[21] Bender, *Angola under the Portuguese*, 106.

[22] Ibid. 104.

colonial cotton became essential to Portugal's competitive success abroad thanks to its low price. In 1947 the price of colonial cotton was 34 per cent lower than the price of international cotton. By 1950 colonial cotton was 50 per cent lower than international cotton. Thereafter, the differences began to diminish slowly as the international price of cotton dropped (see Table A7).

While a low price was necessary, without improvements in quality and quantity, it was insufficient. The war years closed with textile manufacturers and other interested parties expressing doubts about the ability of colonial cotton to satisfy quantitatively and qualitatively the needs of the industry in the post-war years. Now locked into the system of colonial cotton buying and preparing to launch into competitive European markets, importers feared the competition from other countries of textiles produced with better quality cotton. They declared, 'it is urgent to maintain uniform types of cotton so that their value and reputation are increased and they can continue to compete with foreign fibres'.[23] Complaints abounded: fibres were too long, too short, or too varied; too humid or too dry; dirty and full of seeds; and badly ginned and baled.

> The majority of colonial cotton which is traded has such irregular character that a good classifier nearly always finds appreciable differences in the characteristics of two samples taken from the same bale or lot, obtained in the same place.[24]

These worries were only exacerbated by repeated supply shortages and persistent poor quality in the immediate post-war years. In 1947 there was a huge drop in the supply of colonial cotton and Brazilian cotton had to be imported. The GNIAR referred to the problem again in 1948 and 1950.[25] By 1950 textile manufacturers joined the GNIAR in their complaints and worries about the availability of cotton given their expanding needs. They referred to the lack of supply and the poor quality of colonial cotton when available. During the 1950s, the Alcobaça, Rio Vizela, and Cávado companies repeatedly cited the shortage of colonial cotton as a major contributor to the industry's crisis.[26] Part of the reason why colonial shortages so crippled the industry is that their occurrence

[23] Bastos, *O algodão no comércio*, 84.
[24] Ribeiro, *Da colheita*, 88.
[25] GNIAR, *Relatório*, 1948, 1950.
[26] See company reports for the 1950s.

forced the industry to buy more expensive cotton on the international market. With the industry struggling to modernize and gain a foothold in foreign markets, such purchases undercut its ability to compete and revealed how critical to its prosperity was the low price of colonial cotton after the war. The industry was therefore quite interested in improvements in quality and quantity in order to take full advantage of colonial cotton's low price—an advantage which would disappear in the late 1950s. By 1955 the price of colonial cotton was only 2.87 escudos per kilogram cheaper than the international price; by 1958 it was 1.48 escudos more expensive, sending manufacturers into a panic (Table A7).

Notwithstanding the weather, which played its part in erratic output, a number of factors affected the quality and quantity of cotton after the war. With regard to the production process, many Africans worked in destitute conditions. As one observer remarked, 'Africans grow cotton yet continue to walk naked.'[27] To grow cotton, Africans employed inappropriate cultivation methods forced on them by the concessionary companies and frequently in areas which were unsuitable. Companies had little control over individual producers and had inadequate transport. Gins were second-hand and bales irregular in size and weight; storage facilities were damp and dirty.

In Angola, however, one company was convinced that the slothful behaviour of Africans was responsible for the successive decrease in production from 1943. In its annual report for 1946, the Companhia Geral dos Algodões de Angola (Cotonang) stated:

> In the zones in which we work, there are many Africans who carry out the instructions received from our agents on the choice and preparation of land and cultivating operations, but unhappily, the majority of Africans because of their indolence, lack of foresight, and reduced needs, do not follow the instructions, resulting naturally in yields which are very inferior to those which they could easily manage.[28]

Cotonang's statistics for the years 1943 to 1946 in Table 9.1 clearly reveal not only a drop in production but also a drop in the number of cultivators. Though Cotonang did not consider it, the disparity between the high number of cultivators and the low production rate in 1946 would seem to suggest either that

[27] *BGU* 26: 298 (Apr. 1950), 152.
[28] Cotonang, *Relatório*, 1946, 6.

TABLE 9.1. *Cotonang: Number of Cultivators, Areas Cultivated, and Total Production, 1943–1946*

Year	Cultivators	Area (ha.)	Production (tons)
1943	43,881	27,580	12,351
1944	41,843	25,052	8,633
1945	39,052	22,896	7,560
1946	36,644	21,074	6,955

Source: Cotonang, *Relatório*, 1946.

clandestine trading was taking place or that Africans registered as cotton cultivators but were growing other crops.

While few cotton analysts of the period denied the 'laziness' of the African as a possible reason for the drop in production and the poor quality in Cotonang, other explanations were offered and are more plausible. Too much attention was being paid to cotton and not enough to food crops, therefore people were leaving the area, finding other jobs, resisting, or were simply too weak to cultivate.

In addition to poor methods of production, average prices paid to producers in Mozambique for 1947 were 1$50 and 1$10 escudos per kilo respectively (Table A9). Since price differences between first and second quality cotton were not that great, producers had insufficient interest in improving the quality of their yields.[29] Moreover, it appeared that official producer prices, however minuscule the differences between categories, were not being followed in practice. In reality what was occurring was that

> If the cotton is bought in seed it is paid an average price by weight, without a difference being made between diverse grades of cotton which are only recognizable after ginning. Such a system has the inconvenience of disinteresting the farmer in the progressive improvement of quality.[30]

Price also affected quantity in the post-war period. On the one hand, the producer price was too low to encourage Africans to increase output; on the other hand, the difference between the world price and the colonial export price was leading to erratic

[29] Bravo, *A cultura algodoeira*, 179.
[30] Ribeiro, *Da colheita*, 106.

output or clandestine trade. Regarding the former, given that average output per producer in 1947 was still only 105 kilos and assuming it was all 1st quality, at most, the average producer was making 158 escudos per year—an amount hardly worthy of the effort.[31]

As for the oscillation of output depending on the price, the GNIAR complained in 1948:

> In the years in which the official fixed price for metropolitan consumption exceeds or closely follows the quotation on foreign markets, production tends to increase and in the years in which the reverse is true, when the prices of overseas cotton are lower than international prices, production tends to drop.[32]

The GNIAR suggested that some kind of subterfuge was going on but then concluded that this was impossible as both the companies and the government were properly fulfilling their duties. That companies might force Africans to switch to another crop, or cheat and sell cotton on the world market when great differences existed between the colonial export price and the world price were not considered or investigated. Also, colonial administrators and company officials did not recognize Africans' abilities to switch to other crops when the cotton price was low, to engage in clandestine trading to obtain a better price, or to resist growing altogether, regardless of price.

Since the difficulties with cotton began with the African producers, it was to them that the government directed its solutions after the war. These solutions, which ranged from social engineering to the introduction of a notion of status, indicate that the state was willing and able to pursue alternatives to coercion in order to maintain its objectives with regard to colonial cotton production. Secondly, the policy changes suggest a certain responsiveness by the regime to deficiencies in the previous system, and to criticisms and hardships experienced by Africans. In that sense, Africans played some role in shaping cotton policy even if these changes were not always reflected in practice. Below we examine how the regime addressed abuses in the cotton system in the period up to 1958.

In 1945 after a speech by the Colonial Minister, Marcello Caetano,

[31] Bravo, *A cultura algodoeira*, 131.
[32] GNIAR, *Relatório*, 1948, 2.

criticizing forced labour,[33] the state haltingly shifted from a policy based on coercion to one based on the commoditization of households engaged in cotton production. Africans were told that they would not have to work for any other company if they cultivated 1 hectare each of cotton and of food. This was followed by Decree-Law 35,844 in 1946 (the same year as the modernization circular for textiles) which exclusively addressed the entire organization of the cotton regime.[34] The law was both a restatement and a revision of the 1926 decree passed twenty years before. It attempted to address four major problems within the cotton regime: the risks solely undertaken by Africans in cultivating cotton, the high emigration from cotton-growing areas, the abandonment of food crops by those engaged in the cotton regime, and the cultivation of cotton in unsuitable areas. It proposed a new regime in Angola and Mozambique whereby the risks to Africans would be reduced through the elimination of unsuitable areas and through greater technical support by concessionary companies and the JEAC. Only certain zones would be obliged to grow cotton[35] and labour could not be recruited in these areas for any other purpose or by any other company.[36] The legislation obligated companies to verify that Africans grew food as well as cotton and that crops were rotated to replenish the soil.[37] It also exhorted administrators and overseers to take more care, to disseminate better information. The JEAC increased its testing of seeds and conducted soil surveys in order to identify suitable areas and eliminate unproductive ones.

More attention was paid to the process of cultivation and ginning as well. The JEAC advocated the adoption of a single type of press for baling and recommended cotton-harvesting machinery in order to increase the area devoted to cotton and to speed up harvesting time. As mechanization required large areas and inputs of capital, the suggestion was repeatedly made and just as often ignored.[38]

Mechanization might also have contributed to solving the labour problems in both colonies, but instead emigration and conflicts were to be avoided through social engineering: settling Africans in designated areas and providing them with housing, land, and

[33] Vail and White, *Capitalism and Colonialism*, 302.

[34] *Diário do Governo*, Série I, 31 Aug. 1946, D-L 35,844.

[35] Ibid. Section I, Art. 3.

[36] Ibid. Section I, Art. 2, clause 1.

[37] Ibid. Art. 4.

[38] Ribeiro, *Da colheita*, 124.

assistance in order to get them to produce. Long before *aldeamentos* (fortified villages) were constructed to control the population during the colonial wars, the JEAC was establishing cotton *concentrações* (model villages) to improve cotton production.[39] The *concentrações* were supposed to improve the quality of life for Africans by providing them with machinery, teaching them more rational production methods, and allowing the cultivation of food crops. In Angola four *concentrações* were established in Quissama, Ambrizete, Benguela, and Huambo with the guarantee of land, technical assistance, free seeds, free medical assistance, and bonuses if Africans would produce cotton. One hundred and eighty-two families were said to be already settled in these areas. In Catete, the administration planned a *concentração* for 200 families in the zone of Lagos e Irmão, another cotton company. At completion, each family was to have 5 hectares, 3 hectares for cotton and 2 for food; the company would supply tractors and other equipment. In Mozambique, too, the administration envisioned 35 cotton *concentrações* in Niassa.[40]

In 1948, as a part of the *concentração* scheme, the JEAC introduced a two-tiered classification system differentiating cotton producers according to the amount of land devoted to cotton and according to gender. The system went beyond that delineated in Decree-Law 35,844 and tried to introduce the notion of status into African cotton production in the hope that this would attract more people to the regime. Cotton *agricultores* (farmers) would be men, aged anywhere from 18 to 55 years, who farmed one hectare of cotton plus one-half hectare of cotton for each wife beyond the first wife (thus acknowledging that the first wife was, in fact, farming the one hectare cotton *machamba*). *Cultivadores* (cultivators) would be single, widowed, or divorced women or old men between 56 and 60 years who farmed one-half hectare of cotton and an equal area of food crops.[41] In spite of the fact that women were the majority of cotton producers in both colonies, *agricultores*, not *cultivadores*, would be offered bonuses, medical and technical assistance, and the choice of growing area.[42]

[39] Newitt, *Portugal in Africa*, 123, says that model villages were considered as early as the 1930s.

[40] *BGU* 26: 298 (Apr. 1950), 133 and 157.

[41] Bravo, *A cultura algodoeira*, refers to the criteria established by the JEAC in 1948, *Bases para a companha algodoeira*, 117.

[42] Portugal, INE, *I Congresso dos Economistas Portuguesas (problemas das economias ultramarinas)* (Lisbon: Centro de Estudos Económicos, INE, 1955), 137.

To provide additional incentives, Decree-Law 37,523 of 1949 exempted concessions and the few European producers from all taxes, reduced the tax on cotton exports, and forbade deductions from the price paid to African producers for unginned cotton in order to try and increase cotton production.[43] Previously cotton concessionary companies were allowed to take a percentage of the price paid to producers and contribute it to a general development fund for Africans. This meant that the price paid to producers was even lower than officially stated. Following a drop in production in 1951, the Colonial Minister, the Minister of the Economy and the JEAC agreed on a further price rise for producers, managed administratively by the Governors-general of each colony. In both colonies, the value of first quality unseeded cotton was raised by 90 centavos and second quality by 60 centavos in the hope that overall unseeded cotton production would climb to 95,000 tons. At a ratio of three to one, this would have left approximately 32,000 tons of ginned cotton for export to the textile industry in Portugal. It was hoped that these measures along with the price rises would stem the fall in production and interest African growers.[44]

To compensate for the rise in producer prices as a result of the elimination of deductions on unginned cotton, it was promised that the price of ginned cotton exported to the metropole would be increased also, which it was—from 13$50 per kilo in 1949 to 16$30 per kilo in 1951. The intention of the increase in the export price was to appease concessionary companies who had been whining about the low price of colonial cotton in relation to the international price.[45]

The price increases, bonuses, *concentrações*, the introduction of *agricultor* status testify to a regime attempting to replace force with an incentive system based on moral and material rewards. The switch indicates a realization that blatant and visible coercion did not satisfy the quantitative and qualitative demands of the industry, provoked resistance, and was internationally unacceptable—at a time when Portugal was turning towards Europe. To a state newly converted to rationalization, modernization, and other technocratic paraphernalia, coercion was anachronistic and worse, unprofitable.

[43] *BGC* 25: 290–1 (Aug.–Sept. 1949), 216–18.
[44] *BGC* 25: 286 (Apr. 1949), 112–13.
[45] Bravo, *A cultura algodoeira*, 190.

IMPACT OF COTTON CHANGES

Like so many of the regime's policies after the war, these changes did not bring the desired results. As Table A4 indicates, the reorganization of the cotton regime in the late 1940s, tax incentives, price rises, and the establishment of *concentrações* did lead to an increase in average colonial cotton exports during the 1950s, especially from Mozambique, but this supply was still erratic and of poor quality. In 1950 the colonies produced 31,966 tons of ginned cotton for export to Portugal and the metropole imported only 1,471 tons from abroad. The following year, however, production was down again, much to the horror of the textile industry. From 1952 productivity did increase but still failed to keep pace with the needs of the growing metropolitan industry. By 1954 total sales to the metropole equalled 43,076 tons, but these had plummeted to 29,255 tons by 1957.

In Angola, though production per hectare increased during the 1950s, total output stagnated.[46] Once again, companies attributed difficulties with the cotton supply to labour shortages and the administration answered their cries. In 1947 a dispatch by the government in Angola sanctioned what Cotonang had already been practising. According to the dispatch, Cotonang was allowed to employ African overseers (*capataz*) to encourage cotton production in its areas of operation—a violation of the spirit of the 1946 law.[47] Cotonang's 1950 annual report, which mentions that the company was working with the administration to increase the number of cultivators, also suggests that coercion was still a part of the cotton system at the turn of the decade.[48]

In the early 1950s, however, a new challenge arose to the cotton regime in Angola—coffee production. As the report of one Angolan cotton company explained, with the growth of coffee, labour competition, rather than shortage, more appropriately described the period:

> The cotton year was poor, not only owing to the lack of rain in all the littoral region, but also because of the lack of help from administrative

[46] Excerpt from *A província de Angola* in *BGU* 21: 363 (Sept. 1955), 275–6.
[47] *BGC* 23: 264, Apenso (June 1947), 3.
[48] Cotonang, *Relatório*, 1950, no page.

authorities. They did not send to us the amount of labour fixed by the colonial Governor; instead 400 workers left for the coffee fields.[49]

As the coffee boom gathered pace, labour was being enticed or dragooned on to coffee plantations. Angolan administrative authorities and probably the metropolitan government were aware of the vital foreign exchange to be earned from coffee exports and did little to stop the flow of labour. With Mozambique producing the majority of colonial cotton, its cultivation in Angola was not given the same attention by the colonial administration. In 1954 Algodoeira Colonial Agrícola of Angola again complained that local administrative authorities had not sent the allotted amount of labour despite efforts to secure it by the company and the JEAC.[50] The Angolan JEAC, neither as zealous nor as powerful as its Mozambican counterpart, was unable to prevent the decrease. For this company, cotton production did not increase until the late 1950s when the coffee boom receded, the company mechanized, and methods of cultivation improved.

In Mozambique, where the commitment to cotton was far greater, the administration resorted to its old standby—force—when confronted with falling output. After a sharp decrease in output in 1951, government officials in Niassa province of Mozambique adopted an extreme position by declaring that 'any male who failed to produce 400 kilos of cotton would be sent to work without remuneration on the European tea estates', a ridiculous figure since average output per producer in Mozambique was 96 kilos that year and only 184 kilos on average for the 1950s.[51] That same year, an appalling famine occurred at Mogovolas. Mogovolas was supposed to be one of the administration's new model villages implemented as part of Portuguese colonial policy under Decree-Law 35,844 of 1946, ironically the same decree which made food-growing compulsory in addition to cotton. The *circumscrições* of Montepuez and Erati in Mozambique also went hungry in 1955 and 1957.

[49] Algodoeira Colonial Agricola, *Relatório*, 1953, no page.

[50] Ibid. 1954, no page.

[51] Isaacman *et al.*, 'Cotton is the Mother of Poverty', 590. For average kilograms per producer, see Bravo, *A cultura algodoeira*, 131. If the administrator had referred to production per hectare (which assumes more than one producer), he may have been more accurate as averages per hectare were around 360 kilograms per hectare in 1957–60 (but those were good years).

The deaths of 3,000 to 4,000 people in Mogovolas and unnamed numbers in Montepuez and Erati suggest that in some areas officials and companies still gave priority to cotton over food in spite of specific legislation.[52]

Yet the attempt to improve the regime, establish commodity producers, and create an incentive-based system, was not mere rhetoric on the part of the administration. The number of producers did drop as a result of government legislation or other demands. In Angola, producers went from 66,997 in 1946 to 57,435 in 1954. In Mozambique the number decreased from 633,304 producers in 1946 to 514,188 in 1954.[53] More importantly, the 1950s did not see the huge fluctuations of the 1940s in hectarage and number of producers indicating that peasants had responded to the state's measures. For some, cotton had become a part of peasant production and they were staying in it. Concurrent with this stability came an increase in output per hectare and per producer. In both colonies, output per hectare was comparable to the rest of Africa by the late 1950s. Output per producer in Mozambique averaged 184 kilos for the 1950s compared with an average of 103 kilos in the 1940s. Moreover, in several areas of Mozambique, some peasants even saw a return that was 25 per cent higher than the 'ideal' average of 450 kilos per hectare desired by the Governor-General in 1951. This higher output was due to greater productivity of the soil, more investment in materials and tools, better transport, greater involvement by the technical and scientific services of the JEAC, and subsequently, a greater commitment by peasants.[54]

Additionally, some companies did adhere to the regulations on food crops. In a self-congratulatory report, Cotonang spoke of its efforts to encourage food production:

> In collaboration with the authorities we have continued to combat the lack of foresight of the 'native', trying to get him to intensify the production of foodstuffs necessary for his sustenance. To this end, we have distributed seeds of a very good variety of manioc as well as various seeds including

[52] A. Castro, *O sistema colonial português em África* (Lisbon: Ed. Caminho, 1980), 287. Conditions under which peasants produced were also ignored, see Hedges and Rocha, 'Moçambique durante o apogeu do colonialismo português', 43.

[53] *Diário do Governo*, Série I, 24 Nov. 1955, D-L 40,405.

[54] Hedges and Rocha, 'Moçambique durante o apogeu do colonialismo português', 34.

corn and peanuts. With the same objective, we are establishing fields of manioc in order to obtain great quantities of seeds to distribute to the 'natives' of the cotton zones.[55]

While the emphasis in the report is on manioc, which is the easiest and least nutritious of food crops to grow, the mention of corn and peanuts suggests the company took seriously the threat of famine. For this company, the distribution of seeds for food crops also seems to have aided their pursuit of increased output. Production in the Cotonang zones rose from 1948, though the company continued to complain about 'lazy' Africans and the lack of cultivators.[56]

While force evidently was still a feature of cotton production, metropolitan government officials persisted with efforts to devise a system based on moral and material rewards, especially as international prices continued to drop in the mid- to late 1950s. Decree-Law 40,405 of 1955 revealed the government's intention to preserve if not increase the cotton supply by guaranteeing prices, renewing concessions, providing incentives, and even allowing the formation of cotton co-operatives by African cotton farmers. Cultivation in cotton zones was to be accompanied by greater planning and the government undertook the responsibility of protecting producers against risks by creating a Cotton Fund. The government also reassured concessionary companies and metropolitan cotton importers that price stability would be maintained and a yearly quota would be set for the total amount of colonial cotton bought by the metropole. Price stability benefited the metropolitan industry in the years in which international prices were high, and cotton concessionary companies in the years when international prices dropped. A yearly quota assured producers and companies placement in the metropole. Any amount which exceeded the total quota designated for Portugal could be sold abroad.[57]

The decree furthermore encouraged the creation of co-operatives by Africans. Whether African cotton producers were allowed to form co-operatives prior to 1955 is unclear but Decree-Law 40,405 of 1955 officially legislated their inclusion in the co-operative scheme. The role of the cotton co-operative societies would be to acquire, store, gin, and sell cotton on behalf of their members.

[55] Cotonang, *Relatório*, 1946, 6.
[56] Ibid. 1949, 1950.
[57] *Diário do Governo*, Série I, 24 Nov. 1955, D-L 40,405.

Organization of a co-operative had to be approved by the governor of each district (province prior to the 1951 constitutional change) and at least 100 African farmers had to belong before approval was granted.[58]

After thirty years of clinging to a system of forced labour it seems remarkable that the government allowed co-operatives. Combined with the incentive-based system, it is clear that the Portuguese government was liberalizing the cotton regime in order to improve production and to satisfy international opinion at a time when the metropole was looking increasingly towards Europe. Labour shortages in Angola as a result of the coffee boom and erratic yields in Mozambique during the 1950s highlighted the need to improve further the stability and productivity of labour in order to bolster the supply and quality of cotton. The decree even stated that it saw co-operative societies as a 'useful method of collaboration' and that the central objective of the decree as with Decree-Law 35,844 of 1946 which it replaced was to improve production. Additionally, having recently joined the United Nations, Portugal was interested in presenting a favourable impression of its treatment of labour by encouraging voluntary associations of labour, even if, in practice, it continued to operate very much as before.[59]

Labour competition, migration, and resistance revealed the need to find the cheapest, most manageable means of encouraging Africans to produce. Other colonial powers had found that Africans produced best if they were settled on their own plots of land or if they had incentives to produce. Portugal reached a similar conclusion. One approach was the *concentração*; the other, the co-operative. As early as 1947, Belgium had tried the co-operative system in the Congo with good results, so the Portuguese followed, just as earlier they copied the Belgian system of forced labour.[60] Moreover, there was always the possibility that if the total quota for the metropole were exceeded, cotton could be sold abroad to earn foreign exchange for the colonies. The formation of co-operatives then, was really another attempt to increase and improve cotton production.

[58] Ibid.

[59] Revd Clifford J. Parsons, 'The Torment of Angola', *Africa South in Exile*, 5: 4 (July–Sept. 1961), 77.

[60] For the reference to co-operatives in the Belgian Congo, see *BGC* 23: 263 (May 1947), 45.

At least one African cotton co-operative formed after the passage of the 1955 legislation. In the district of Cabo Delgado, northern Mozambique, 12 African farmers started the Mozambique African Voluntary Cotton Society in 1957. Within two years it had over 3,000 members and more importantly, it had achieved a higher per capita cotton output than non-members of the co-op.[61] Its formation was initially supported by the JEAC and SAGAL (the cotton concessionary company responsible for cotton production and ginning in that area) both of whom were anxious to prevent the loss of labour to neighbouring sisal plantations.[62] In addition to the higher per capita cotton yields, food production on the lands of co-operative farmers also increased. Finally, co-op members were able to negotiate with SAGAL a 40 per cent increase in the price paid to producers for first quality unseeded cotton, as an incentive to produce better quality cotton. Thus the formation of a cotton co-operative appeared to be solving several of the problems which had worried the JEAC and companies in the past: quantity, labour shortages, food production, and possibly quality. Yields and food production had increased; mechanisms were now in place to reduce labour shortages and improve the quality of cotton.

In the rest of the cotton-producing areas, the JEAC knew from past experience that increases in production and improvements in quality followed increases in prices so producer prices were also raised in 1955. There was also much talk of mechanization but little seems to have occurred in the 1950s. In addition experiments with Egyptian strains were reportedly begun in Mozambique to try and produce a longer fibred, better quality cotton, greatly desired by the domestic textile industry.[63]

After low returns in 1956 and 1957, production was quite high in 1958 (Table A4). Yet international events foiled the regime's efforts once again. By 1958 the colonial cotton price was higher than the international price of cotton. In spite of all of its alterations, the regime had not outpaced technological advances in other countries which had produced drops in the price of cotton. Naturally,

[61] Allen Isaacman, 'The Mozambique Cotton Cooperative: The Creation of a Grassroots Alternative to Forced Commodity Production', *African Studies Review*, 25: 2–3 (June–Sept. 1982), 5.

[62] Ibid. 13.

[63] Mario de Carvalho, 'Possibilidade de cultivo de algodões do tipo egípcio em Moçambique', *Boletim da Sociedade de Estudos da Província de Moçambique*, 27: 111 (July–Aug. 1958).

the Portuguese textile industry balked at the prospect of buying from the colonies and criticized colonial prices, but the government maintained the colonial price at its old level. After 1958 the decision to maintain the colonial cotton price would bring severe repercussions in the colonies and at home.

THE REVIVAL OF POLITICAL DISCONTENT

The regime's experiment with incentives and co-operatives in the cotton scheme did not extend to other sectors of the economy and were not applauded by all. Economic improvements in the colonies conformed to priorities determined by the metropole and were not accompanied by significant political reforms. For example, most industries were owned by metropolitan or foreign companies and organized and run according to their own requirements. Settlers and Africans in the colonies lacked capital but more importantly, were not allowed to make decisions which affected their lives or the places where they lived. Bureaucratic posts and political appointees in the colonies performed mostly administrative tasks; political and economic decisions were taken mainly in Lisbon.

In the 1950s, settlers frequently voiced discontent with the pattern of economic development in the colonies. From labour shortages to the lack of industrialization, settlers felt that the metropole was taking advantage of the colonies. They criticized the Industrial Regulations for preventing industrialization and favouring special interests. As one writer for the *BGC* put it:

> The monopolies and oligopolies restrict production, reduce supply, affect prices, and destroy precious goods. They fail to satisfy the immense needs of humans and spread materialistic evils through society.[64]

Imports from the metropole were overpriced (as we witnessed in the case of textiles) and the percentage of colonial exports to Portugal was dropping. Even after the Development Plan of 1953–8 proposed loans, improvements in transport, communications and infrastructure, settlers criticized Portugal's slow colonial industrialization and the government's support for certain companies. An article in the *BGU* (formerly *BGC*) reprinted from a

[64] *BGC* 26: 308 (Feb. 1951), 91.

Mozambican newspaper charged that Industrial Regulations were holding back the colonies and that the metropolitan and colonial economies were badly co-ordinated. It also argued that the colonies lacked capital and industry and that energy costs were high.[65] Back in Portugal, Caetano thought it necessary to defend the Portuguese government against charges that its representatives were found on company boards; that metropolitan companies were making excessive profits in the colonies; and that the standard of living was low. Whether the charges were just or not, the fact is that the Portuguese government had made emigration one of the cornerstones of its colonial policy since the war but was then unprepared to accommodate settler demands.

While settlers voiced their discontent through official organizations, Africans in Mozambique and Angola as well as those in exile also formed groups to bring political change to the Portuguese colonies. These ranged from reform-minded groups such as the Associação dos Naturais de Moçambique (Association of Mozambican Born) who wanted an end to forced labour and other abuses to more radical groups such as the MPLA (Movimento Popular de Liberatação de Angola) in Angola, founded in 1956, who wanted independence and revolutionary social change. Though they had little influence with the Portuguese government at this time, these groups were able to attract support from independence groups in other African nations and to arouse international awareness of Portuguese activities in the colonies. After Portugal's admittance in 1955, the United Nations began inquiries into the legitimacy of Portugal's empire in the light of international law.[66] In the late 1950s, the International Labour Organization also launched an investigation of Portugal's observance of the Forced Labour Convention of 1957.[67]

As in the metropole then, the government's colonial policy still contained severe flaws by the late 1950s. While Portugal appeared to be laying the groundwork for economic development, these efforts were piecemeal and not matched by similar political initia-

[65] *BGU* 31: 355 (Jan. 1955), 394–6.

[66] George Martelli, 'The Issues Internationalized', in Abshire and Samuels (eds.), *Portuguese Africa*, 375–85.

[67] International Labour Organisation, *Portugal in Africa*, Report of ILO Commission on the Observance by Portugal of the Abolition of Forced Labour Convention, 1957 (no. 105), Findings and Recommendations (London: Portuguese Embassy, 1962).

tives. The government neither built political support amongst settlers nor accommodated African demands for basic human rights, much less revolutionary change. Policies remained directed from above and were often poorly implemented and contradictory.

With regard to cotton, like much of the regime's post-war colonial legislation, changes in policy were not fully effective in practice. Force was still a feature of the regime and heavy administrative involvement continued in the post-war period when the metropolitan textile industry needed the cotton. Nevertheless, the picture of cotton policy which emerges after 1946 contrasts sharply with the image of Portuguese rule in the colonies so often presented of a regime blindly, stubbornly, and anachronistically clinging to forced labour. Rather, the evidence suggests that the regime was trying to abandon force in favour of material and moral rewards in order to obtain a better and more stable supply of cotton. Such action mimicked the technocratic, developmentalist orientation of policy in the metropole, but unhappily so did the results.

10

The Decline of the Authoritarian Regime, 1958–1974

INTRODUCTION

In analyses of the period from 1958 to 1974, two views tend to coexist uneasily in the literature; this chapter intends to reconcile them. The first view is a theory of stagnation: an interpretation that, economically and politically, the *Estado Novo* stood still during the period. It did not further institutionalize the corporative principles embodied in the Constitution but neither did it move substantially in the opposite direction towards genuine liberalization. In this interpretation, the regime is seen as isolated and unchanging, and its policies, static and chaotic. The Portuguese economy is depicted as still primarily agricultural with a weak and slow development of industry, and as dependent on the colonies for economic survival.[1]

The other interpretation, the integrationist theory sees the period as one of great change and development with the regime acting on behalf of groups of financial-industrial conglomerates which have emerged during the post-war period. This view documents extensively Portugal's growing integration with Europe in the areas of trade and investment and contrasts it with Portugal's waning colonial economic connection to argue that the metropole did not need, or had no interest in the colonies by 1974. Although the colonial wars first evoke a sense of nationalism on the part of metropolitan groups, the wars are later viewed as a hindrance to

[1] See Gallagher, *Portugal*; Wiarda, *Corporatism and Development*; Pinto de Almeida, *A indústria portuguesa e o condicionalismo institucional* (Lisbon: Seara Nova, 1961). For Portugal's dependence on the colonies, see Anderson, 'Portugal and the End of Ultra-Colonialism'; Jay O'Brien, 'Portugal and Africa: A Dying Imperialism', *Monthly Review*, 26: 1 (May 1974), 19–36; and Ali A. Mazrui and Michael Tidy, *Nationalism and New States in Africa* (London: Heinemann, 1984), ch. 9.

further integration with Europe.[2] Both interpretations see the regime jettisoned by its former supporters by 1974 in order to make way for a regime which can more effectively address new political and economic demands.

Both of these views contain valid observations on Portuguese politics in the 1960s and early 1970s, yet some combination of the two interpretations would present a more accurate and less paradoxical picture of events. State policy was certainly arbitrary and inconsistent in this period but in continuation with the previous period the state did support industrialization and made substantial efforts to increase export opportunities for many industries. In contrast to the interpretation presented by many stagnation theorists, economic policy was quite dynamic (though political evolution was not); the most notable aspects were growing integration with Europe and foreign investment in Portugal during the 1960s and early 1970s. On the other hand, integration with Europe did not exclude a continuing interest in Africa as many integrationists seem to imply. One could invest both in Europe and in the colonies and indeed, this was the policy of both the *Estado Novo* and several of the large financial and industrial groups as we shall see. Other Portuguese industries (including textiles), moreover, continued to depend on the colonies for raw material supplies and markets and were not fully integrated into European markets at the start of the wars. So long as Portuguese capital could operate in Europe and Africa, the two policies worked in conjunction with each other, though the balance was changing.

With the continuation of colonial violence during the 1960s, such a partnership became less and less tenable. Colonial resistance, international pressures, and the economic and political costs of fighting wars in the colonies mounted. Coupled with the resurgence of old internal problems, the war weakened the position of the government *vis-à-vis* the opposition, and fragmented powerful economic élites in Portuguese society subsequently diluting their support for the regime. These dominant groups and elements in the

[2] Vaz, 'Portugal and Her Overseas Territories'; and Clarence-Smith, *The third Portuguese empire*; Rafael *et al.*, *Capitalismo*; and Robin Blackburn, 'The Test in Portugal', *New Left Review*, nos. 87–8 (Sept.–Dec. 1974), 5–46, lean towards the integrationist interpretation to explain the downfall of the *Estado Novo*, but allow for continuing economic links in the colonies amongst the financial-industrial conglomerates. Neither investigates in much detail the political reasons for remaining in the colonies.

Portuguese army linked to leftist opposition circles respectively reconsidered their positions and began to argue for a neo-colonial solution. The political requirements and bases of support of the regime, reinforced by remaining colonial economic links, blinded it to such a solution, which prolonged the war and ultimately contributed to the regime's collapse.

While the colonial wars were an area of contention in the 1960s, I would like to suggest that economic and political difficulties which had arisen by 1958 remained an area of conflict during this time too. The colonial wars accentuated the difficulties and accelerated certain trends which had already become visible by the early 1960s. It was the government's failure to solve several of these problems combined with the colonial wars which led to the downfall of the authoritarian regime in 1974.

THE OPPOSITION AND THE 1958 PRESIDENTIAL ELECTION

While the government faced charges of incompetence, intervention, and inconsistency from part of the economic élite by the late 1950s, political opposition grew domestically, colonially, and internationally. The opposition criticized low wages, the poor standard of living, the unfair control of many economic sectors by monopolies, and state favouritism. Equally, protests focused attention on the lack of civil liberties, unfair electoral methods, repression, censorship, and torture in Portugal and the colonies. Much of the population was deprived of political power and excluded from a share of economic rewards.[3]

The extent of opposition had been evident in the 1949 and 1951 elections, but it was the 1958 presidential election which provided the greatest domestic shock for the government. Though the opposition had been effectively divided during the 1950s over

[3] For opposition and Delgado's campaign in the late 1950s, see Herminio Martins, 'Opposition in Portugal', *Government and Opposition*, 4: 2 (Spring 1969), 250–63; relevant chapters in Raby, *Fascism and Resistance*; and Cruz, *Monárquicos e republicanos*. For a detailed discussion of the armed forces' role in the maintenance as well as the downfall of the authoritarian regime (which falls outside the scope of this book), please see Antonio Rangel Bandeira, 'The Portuguese Armed Forces Movement: Historical Antecedents, Professional Demands, and Class Conflict', *Politics and Society*, 6 (1976), 1–56.

the issue of whether to participate in or abstain from elections, it united behind the candidature of General Humberto Delgado in 1958. Delgado had participated in the 28th of May coup in 1926, contributed to the establishment of the *Estado Novo*, and represented Portugal in NATO, but had become disillusioned with the regime. Backed by groups ranging from the Communists to the Monarchists and even sections of the armed forces, Delgado challenged Admiral Tomás, the official candidate for the Presidency.

Delgado's platform was a moderate one. He pledged to end corruption, guarantee public order within the law, amnesty political prisoners, improve the standard of living, hold free and fair elections, uphold international treaties including Portugal's NATO alliance, and enforce the Constitution of 1933 (not 1951) because of its references to fundamental civil rights. Most importantly, if elected, he promised to exercise his constitutional right to discharge the Prime Minister and to choose a new government more responsive to the wishes and needs of Portuguese people.[4]

Delgado received substantial support. Not only had he been previously a faithful servant of the regime but also his platform had not called for violent overthrow of, only adherence to, the Constitution. Threatened by this criticism of its legitimacy and the popular support he roused, the regime responded with repression. Publications were censored, Delgado's house attacked, opposition members imprisoned and murdered, and rallies broken up by the police. Not surprisingly, Delgado lost the election but 'never had the foundations of the regime been so badly shaken'.[5] Despite ballot rigging and intimidation, Delgado did better than any previous opposition candidate, receiving a quarter of the vote in Continental Portugal.[6] Additionally, colonial settlers fed up with Lisbon's interference strongly backed Delgado in Mozambique and Angola (see Ch. 12). To prevent any future electoral challenges to the regime, following the election a constitutional amendment provided for the indirect selection of the President by an electoral college.[7]

Political opposition, the conduct and outcome of the election,

[4] Cruz, *Monárquicos e republicanos*, 144–5.

[5] Ibid. 150.

[6] Bailey, 'The Political Process' in Abshire and Samuels (eds.), *Portuguese Africa*, 153.

[7] 'The Portuguese Constitution', in Peaslee, *Constitutions of Nations*, ch. 2, sect. 1, Art. 72 (1959).

along with criticism by technocrats, representatives of large conglomerates and of traditional industries suggest that the regime was under acute pressure, domestically and internationally, as the decade ended. Several economic groups in Portugal wanted a greater commitment to industrialization, integration with Europe, and less arbitrary and inconsistent treatment by the regime. In addition to criticisms of the authoritarian nature of the regime, came more serious charges by the United Nations about Portuguese labour practices in the colonies and its denial of their right to self-determination. In the colonies themselves, mounting African discontent and settler complaints confronted the government. The most visible sign of the displeasure was in Goa to which Nehru devoted most of his attention from 1953 on. In the event, Salazar was more effective at outflanking and repressing his political opponents than at resolving the underlying conflicts and problems in the political system and the economy.

Following the humiliation of the 1958 election, the regime returned to its economic difficulties. In the previous chapter on the regime, we witnessed the emergence of contradictions in the state's political and industrial policies. By 1958 the increasingly technocratic and *laissez-faire* approach adopted by industrialists and some government members was beginning to lead to conflicts with those resistant, or unable to change. The technocratic approach, originally propagated to consolidate the regime and to deflect accusations of authoritarianism after World War II, now threatened to divide it.[8] The government willingly supported the growth of monopolies but failed to anticipate the pace of their expansion, their interest in moving beyond the domestic market, their ability to control key sectors of the economy, or the consequences of their growth on the traditional sector. This resulted in dissatisfaction among not only those who had been aided by the regime and now wanted to move on, but also those who had been overlooked or stifled by the regime, even as the government attempted to meet the complaints with promises to modernize and to conclude trade agreements. According to Elisabeth Leeds:

[8] Portugal, Assembleia Nacional, *II Plano*, iii, see esp. the statements by Sr Deputado José Sarmento, 36–44; Nunes Mexia, 45–83; João Dias Rosas (the ex-President of the CRCAR), 135–51; Araujo Correia, 221–57; and Proença Duarte, 336–60. In addition, Ferreira Dias, the Under-Secretary of State for Commerce and Industry from 1940 to 1944 and a major supporter of the technocratic approach was back as Minister of the Economy from 1958 to 1962.

By 1960, a growing body of technocrats, following in the footsteps of Ferreira Dias and the more progressive of the large industrialists, looked to the joining of EFTA as a badly needed stimulus to an industrial economy they recognized as lagging far behind the rest of Europe.[9]

By contrast, the regime's slavish devotion to financial stability, intervention, Industrial Regulation, and protectionism meant that many traditional industries had been unable or unwilling to modernize. By the late 1950s then, with the exception of a minority of modern firms in these sectors, most traditional industries were in crisis and had to rely solely on low wages to remain competitive at home and abroad. Persistent low wages completed the vicious circle by limiting domestic consumption, thereby reducing the attractiveness of the internal market for industry. Worse, low wages meant social disruption, expressed through political opposition or emigration. By the late 1950s, critics of the regime called for substantial changes in the state's industrial strategy.

The regime attempted to accommodate criticisms and overcome the crisis through policy changes. Financed by both public and private investment, the Second Development Plan of 1959 to 1964 was more far-reaching and more specific than the First Plan in its emphasis on industrialization and modernization. According to Section VII:

> The plan will give preference to the installation of industries which, established on solid technical and economic foundations, assure greater employment of labour per unit of capital invested, produce goods destined for export or the substitution of imports from abroad, use national primary materials, are subsidiaries of other existing activities in the country, or contribute to regional development.[10]

The plan allocated 31 million contos (71 per cent of it in the metropole) to be spent on agricultural and industrial improvements, the expansion of electricity, transport, and communications, and the development of research and technical studies. It aimed not only to accelerate the growth of the national product and to modernize, but also to improve the standard of living and reduce unemployment.[11] In the colonies, approximately 9 million contos financed from colonial budgets, loans, and state credit were

[9] Elisabeth Leeds in Bruneau *et al.* (eds.), *Portugal in Development*, 32.
[10] Portugal, Assembleia Nacional, *II Plano*, i. 13.
[11] Ibid. i. 8–10.

designated for improvements in electricity, industry, fishing, communications and transport, education and health, and local projects.[12]

Alongside its more ambitious Second Development Plan, the regime more actively favoured and aided manufacturing and financial capital through trade agreements with Europe and America, foreign investment in Portugal, the supply of credit, state participation in business enterprises, and informal patronage. More medium- and long-term credit was made available through the Development Bank (Banco de Fomento—1959) and the state savings bank, the Caixa Geral dos Depósitos.

The *Estado Novo* concluded specific trade agreements to increase exports, too. In the early 1950s Portugal had trade arrangements with Brazil and other countries, but the real push for exports into Europe came in 1957, the year the Common Market was formed. Since Portugal's major trading partner, Great Britain, had 'ostentatiously stood aside when the Community was founded in 1957',[13] Portugal followed her into the other Western European trading bloc, the European Free Trade Association (EFTA).

By the late 1950s preliminary agreements on membership in EFTA had been concluded with the signing of the Stockholm Accord. Portugal joined the World Bank in 1959, became a full member of the EFTA in 1960, and signed the Gatt Agreement in 1961. It also concluded separate trade agreements with the US, Brazil, and several developing countries during this time.

At the same time, the government responded to the crisis in the traditional sector by creating the National Institute of Industrial Investigation (Instituto Nacional de Investigação Industrial) to provide technical assistance to the industries in crisis including wood, paper, textiles, food-processing, ceramics, and chemicals—again mostly the traditional industries.[14] Reorganization plans were also proposed for sugar, glass, metallurgy, cork, shoes, electrical materials and plastic, and resin.[15] In the event that reorganization

[12] Ibid. i. 15–16.

[13] Dennis Kavanagh, *British Politics: Continuities and Change* (Oxford: Oxford Univ. Press, 1985), 6.

[14] D-L 42,120 and 42,121; *Diário do Governo*, Série I, 23 Jan. 1959.

[15] Portugal, Assembleia Nacional, *II Plano*, i. 86–7; and see Portarias 17,719 (7 May 1960); 17,724 (11 July 1960); 17,827 (15 July 1960); 17,880 (4 Aug. 1960); 18,005 (17 Oct. 1960); and 18,116 (12 Dec. 1960) in *Diário do Governo*, Série I, 1960.

might lead to unemployment, the regime had a twofold response: in accordance with the Second Development Plan, it hoped to absorb labour through greater industrialization, and it retained a controlled policy of emigration. As before, emigration would be used to drain off excess and potentially disruptive labour from the Portuguese economy. It would also be an indirect method to force industries to raise wages, an objective the regime had tried previously but had been unwilling to enforce when industrialists rejected it.[16]

WAR IN THE EMPIRE

Events elsewhere in the Portuguese empire interrupted the changes the regime hoped to achieve with these new policies. During the period 1961–2, isolated skirmishes took place in Portugal's African colonies, but because the regime was under extreme political and economic pressure at home, it overreacted. Small skirmishes were magnified tenfold. The Salazar Government used the threat of a large-scale bloodbath in Africa as a focus for national unity and as a ruse to shift attention away from domestic troubles. By 1962 a prolonged war seemed inevitable.

Certainly there were plenty of reasons why Portugal should want to remain in Africa. Corporatism had been more rhetoric than reality, but nationalism, Portugal and the extended Portuguese nation had been themes to which Salazar had always returned. Africa had been an essential part of this nationalism.[17] In school, children were given a map of Europe, the so-called 'rose-coloured' map, on which the Portuguese colonies were superimposed to convince children that Portugal was really a huge nation. Portugal's imperial greatness and its colonizing mission were constant themes of the official ideology and public monuments. For many Portuguese, it was difficult to imagine Portugal without its African possessions; for Salazar, it was impossible.[18]

Portugal also had invested almost 7 million contos in the colonies

[16] For an analysis of emigration as a conscious policy of the state, see Leeds, 'Labor Export, Development', esp. chs. 5 and 6.

[17] Martins in Woolf (ed.), *European Fascism*, 334.

[18] See Eric N. Baklanoff, *The Economic Transformation of Spain and Portugal* (New York: Praeger, 1978), 106, a 1971 poll shows 57 per cent felt Portugal's future depended on ties with Africa.

from 1953 to 1961 through the Development Plans, increased and promoted white settlement, and continued to strengthen the economic links between the metropole and the colonies. The metropole had also benefited from the healthy balance of payments that the colonies enjoyed with other countries to correct its own chronically negative trade balance.[19] Angola's contribution was mostly due to its foreign exports, including oil and minerals; Mozambique's came from payment by South Africa for the use of port facilities in Lourenço Marques and the employment of Mozambican labour in the South African mines.

Moreover, if the colonies did not bring general economic prosperity to the Portuguese people as a whole, some sectors of Portuguese business and finance clearly prospered from cheap supplies of raw materials, guaranteed markets, and protected investment. Several of the seven identifiable financial-industrial conglomerates which now dominated the Portuguese metropolitan and colonial economies had extensive investments in the colonies. Espírito Santo was probably the most heavily involved, with over 2 million contos worth of investment in various industrial and agricultural enterprises, including the Mozambique company, and sugar and coffee companies in Angola and Mozambique.

The Português do Atlantico group backed the Sociedade Algodoeira de Fomento Colonial which ran Textang in Angola, had a hydroelectric plant in Mozambique, and shares in the Mozambique Company. Other involvements of the Português do Atlantico group in the colonies included banking, brewing, cotton and textile industries, and hydroelectricity. It had interlocking directorships with over seventy colonial companies and directly controlled forty-three others.

Alongside the Anglo-American Corporation and the Société Générale de Belgique, Banco Burnay (in 1967 it became the Banco Fonsecas e Burnay after a merger with Banco Fonsecas, Santos e Viana) had shares in Diamang, the Angolan diamond company (total capital 865,000 contos), listed in 1972 as Portugal's largest exporter, and also owned an important colonial company, Grupo Entreposto, a subsidiary of the Mozambique Company. Grupo Entreposto was involved in approximately twenty different colonial

[19] Eduardo de Sousa Ferreira, *Portuguese Colonialism from South Africa to Europe* (Freiburg: Aktion Dritte Welt, 1972), 28.

activities including automobiles, ceramics, cotton plantations, iron and uranium mining, mostly in Mozambique.

The group organized around the Banco Nacional Ultramarino had most of its interests in the colonies mainly in areas such as maritime transport, tourism, and insurance. It held minority positions in some of the large colonial companies such as Diamang and Cotonang in Angola and the Companhia Colonial do Buzi in Mozambique, and colonial sugar and cellulose firms. It also controlled 20 per cent of the capital in the Companhia Colonial de Navegação, a shipping company. Finally, it had an interest in Secil—the Companhia Cimentos do Ultramar, a cement company.

Borges e Irmão had investments in petroleum in Angola, beer and pharmaceuticals in Mozambique. CUF and the Champalimaud group also had investments though on the whole, they were less dependent on the colonies than the other groups. CUF controlled the oilseed trade in the Cape Verde islands and Guinea, and had 30 to 40 per cent of Guinea's import–export trade. It also had commercial and insurance interests in Angola; and tobacco and sacking operations in Angola and Mozambique. Champalimaud had a monopoly on cement in both colonies.[20]

Many traditional industries were also dependent on the colonies for raw materials and markets. In 1960 exports to the colonies still accounted for 25 per cent of Portuguese exports; textiles and wine, particularly from small and medium producers, accounted for approximately 40 per cent of exports to the colonies in 1960. Companies most assuredly would have been hurt by the loss of these markets given their uncompetitiveness on world markets and the poor quality of much of their production. The remainder of exports to the colonies were accounted for by a diversity of products including cement, glass, beer, olive oil, leather, soap, and pharmaceuticals; again, almost all were associated with traditional industries.[21] As to imports from the colonies, these still accounted for 14 per cent of total imports in 1960, and sugar, cotton, vegetable oils, and oilseeds made up 75 per cent of imports.[22] Were drastic changes to occur, not only the companies which produced and

[20] Information on colonial investments by conglomerates from Clarence-Smith, *The third Portuguese empire*, ch. 6; and Maria Belmira Martins, *Sociedades e grupos em Portugal* (Lisbon: Editorial Estampa, 1973).

[21] Clarence-Smith, *The third Portuguese empire*, 160–2.

[22] Ibid. 151–2.

marketed these materials (themselves financed for the most part by Portuguese capital) but also companies in Portugal would have been severely affected. It is therefore not surprising that some Portuguese wished to remain in Africa in some capacity.

The skirmishes in Portuguese colonial Africa and the regime's violent responses to them provoked international criticism of Portugal's role in the colonies with many critics calling for self-determination, including Brazil and the United States. In Portugal itself, there was general agreement that Portugal should remain in Africa, but it was not a foregone conclusion that a war should be fought to obtain that objective. Amongst the Portuguese opposition, except for the FPLN, an Algiers-based movement of Portuguese exiles who linked the defeat of domestic fascism with independence for the African colonies, most groups, including the Communist Party, envisioned a continuing Portuguese presence in Africa, possibly in a Commonwealth-type organization or in a Lusitanian community.[23] Within the government, even Caetano was said to have favoured a federal solution to the colonial problem.[24] The desire to remain in Africa was not simply a knee-jerk nationalist reaction to the events there but had cultural and economic roots as well.

Although the government's own interest in European integration, trade and investment, and liberalization policies would seem to suggest a neo-colonial solution to the colonial dilemma, a hard line prevailed and a military response adopted. Portugal's position in Africa would remain unchanged. The hard line position appears to have been chosen for the following reasons. The hijacking of the Portuguese ocean liner, the *Santa Maria*, on 22 January 1961 by Henrique Galvão, a popular member of the opposition, occurred just as riots broke out in Angola, and probably explains the regime's initial over-reaction to events there. Moreover, in the midst of the troubles in Africa, an attempted military coup occurred in Portugal which was said to have been seeking a pacific solution to the colonial problem.[25] The coup was crushed, reinforcing the

[23] See speeches by Cunha Leal, Homem de Mello, and H. Galvão in Ronald H. Chilcote, *Emerging Nationalism in Portuguese Africa: Documents* (Stanford, Calif.: Hoover Institution Press, 1972), 28–42.

[24] Cruz, *O Partido e o Estado*, 73.

[25] P. D. Gaitonde, *The Liberation of Goa: A Participant's View of History* (London: C. Hurst and Co., 1987), 127; and Martins, 'Opposition in Portugal', 255.

position of the hard-liners in the government and the military. Following the coup, the abrupt takeover of Goa in December 1961 further hardened the position of the Portuguese government, which was determined that Goa's takeover not be taken as a precedent in Africa. Finally, since the early 1950s, the conservatives in the government had wrestled with the technocrats over economic issues, and Monarchists had seen their political influence gradually eroded. Making the African issue one of life or death, hard-liners and Monarchists used the colonial situation as an opportunity to reassert their power within the regime.[26] By 1962 there were 20,000 troops in Angola and 12,000 in Mozambique and it was clear that a military solution to the troubles in Africa had been adopted.[27] In late 1962 the moderate Minister of the Colonies Adriano Moreira was removed and replaced by a military officer.[28] The wars were used, as wars have been used before and since, to rally initial support for a beleaguered regime. As the wars dragged on, and the outcome became increasingly uncertain, the war sapped support for the regime it had meant to reinforce.

Like the Argentine military government in its 1983 attempt to recapture the Malvinas/Falklands, the Portuguese government calculated that the African wars would provoke an upsurge of nationalism in Portugal. To a certain extent, they did. Even though people were initially unsure about the military option, Salazar received support in the early years of the war. The Monarchists who had withdrawn from the National Assembly and the Corporative Chamber in 1955 rallied behind Salazar and the wars probably allowed him to regain political control after the turbulence of 1958 and 1961.[29] Those groups who were dependent on Africa for markets or raw materials, who were not yet orientated towards Europe, or who had heavy colonial investments were also supportive.[30]

Among the conglomerates, doubts surfaced rather quickly however. Old problems reappeared and the wars introduced a new set

[26] On the hard-liners, see Rafael *et al.*, *Capitalismo*, 99. On the Monarchists, see Cruz, *Monárquicos e Republicanos*, 210.

[27] Troop figures for Angola from UN, Office of Public Information, *A Principle in Torment*, ii: *The United Nations and the Portuguese Administered Territories* (New York: UN, 1970), 17; and for Mozambique, from Munslow, *Mozambique*, 83.

[28] *BGU* 38: 449–50 (Nov.–Dec. 1962), 6–10.

[29] Cruz, *Monárquicos e republicanos*, 210–11.

[30] João Martins Pereira, *Indústria, ideologia e quotidiano* (Oporto: Edicões Afrontamento, 1974), 24.

of difficulties. The wars simply became another focus for dissatisfaction. As early as 1962, the *British Survey* reported that:

It is not surprising . . . to find among realistic Portuguese officials and businessmen, who have no sympathy with emigré conspirators and fully appreciate what Salazar has done for Portugal in the past, a conviction that evolution in the direction of a Lusitanian Commonwealth is the way of salvation.[31]

Alongside the support for a peaceful neo-colonial solution, some financial-industrial conglomerates renewed their criticism of the regime and pushed harder for integration into Europe. Portugal had already joined EFTA in 1960, but was nervously following British attempts to join the EEC knowing that, unless it should gain admittance as well, its trading position would suddenly become precarious. Britain had applied for admission to the EEC in 1961 and Portugal initiated an application but both were turned down. This was little comfort to the financial-industrial conglomerates as the *Africa Institute Bulletin* based in Pretoria reported in 1963:

Recently groups in Portugal who have traditionally opposed Dr. Salazar's regime, have found allies in the wealthier classes who see their interests threatened by a tendency towards state socialism. Their fears also stem partly from Portugal's inability, so far, to gain admittance to the E.E.C. According to these groups, Portugal's association with the E.E.C. offers the only real way out of isolation and the only insurance against the consequences of the ultimate loss of the African provinces.[32]

In a move reminiscent of its behaviour in the early 1930s, the regime tried to balance disparate and conflicting interests and placate the fears of the conglomerates. It tried to satisfy those tied to the colonies and those wishing to integrate with Europe, arguing as late as 1972 that integration with Europe and unity with the colonies were both possible and beneficial.[33] Equally, it sought to meet the needs of major economic groups while maintaining, as always, its own objectives. That these positions produced contradictions and divisions which helped to weaken and then bring down the government, it either overlooked or could not avoid.

[31] 'Portugal in Africa', *British Survey*, no. 158 (May 1962), 13.

[32] 'Portugal in Africa', *Africa Institute Bulletin*, 3: 10 (Nov. 1963), 289.

[33] Augusto de Sá Viana Rebelo, 'Discurso', *A Indústria do Norte*, 123: 635 (Nov. 1972), 16–19.

Below we look at the regime's efforts to accommodate the demands for greater integration and the factors which undermined its position over the years.

In spite of not gaining admission to the EEC, Portugal's membership of EFTA is partly responsible for the growth that Portugal underwent during the 1960s. Imports from EFTA countries increased from 20 to 24 per cent between 1960 and 1970 compared with a decline in the share of imports from the colonies from 14 to 9 per cent in the same period. Exports grew by 8.5 per cent per year between 1960 and 1972, 18.7 per cent per year to EFTA countries. Whereas only 21 per cent of Portuguese exports went to EFTA countries in 1960; 41 per cent of exports went to EFTA countries in 1970 compared with only 15 per cent of exports going to the colonies.[34] Due to the growth of exports and the domestic market, the value added of most industries doubled or even tripled in the period between 1960 and 1973.[35] With 1953 as the base year, the overall index of industrial production was 205 in 1968.[36]

Throughout the 1960s, the regime tried to match the role of EFTA and retain the support of the economic élites by arranging credit, promising further development, liberalizing Industrial Regulations, promoting exports, and attracting foreign investment. From 1962 restrictions on borrowing abroad were relaxed and credit from the United States, Germany, and the World Bank became available to Portugal. From 1965 links with foreign business were encouraged as the Portuguese government dropped most Industrial Regulation laws and invited foreign investors into Portugal. Between 1961 and 1967, foreign long-term capital equalled around 20 million contos, ten times what it had been from 1943 to 1960.[37] In the short period from 1965 to 1966, 20 per cent of investment in the private sector was foreign,[38] but only in 1971 did foreign investment per year in Portugal pass the 1 million conto mark.[39] That year, total investment equalled 1.5 million contos.

[34] Baklanoff, *Economic Transformation*, 133.

[35] Ibid. 126.

[36] Vasco Vieira de Almeida, 'Changes in the Structure of the Portuguese Economy', *EFTA Bulletin*, 10: 8 (Nov. 1969), 14.

[37] Joaquim Ramos Silva, 'As relações económicas Luso-Americanas e a adesão de Portugal a CEE', unpub. paper presented at conference of the Instituto de Estudos Estratégicos e Internacionias, Jan. 1989, 28.

[38] Rafael *et al.*, *Capitalismo*, 61.

[39] Silva, 'As relações', 28.

Businesses in Portugal and the colonies welcomed the foreign capital, excluded for nearly forty years.

The 1965 to 1967 Interim Development Plan devoted nearly 35 million contos to development in the metropole, 14 million of it to industry.[40] The Third Development Plan from 1968 to 1973 allocated approximately 122 million contos over five years in order to promote growth and modernization, to facilitate a more equitable distribution of the country's wealth, and to correct regional infrastructural, agricultural, and industrial differences. Much of the plan was orientated towards improving the ability of Portuguese industry to compete more effectively in Europe and to modernizing the Portuguese economy so that it approximated those in the rest of Europe. The plan gave nearly 31 million contos to industry, 27 million to transport and communications, 17 million to energy, and nearly 15 million to agriculture.[41]

These measures were circumscribed by the regime's own financial and political priorities as so often had occurred in the past. As one observer wearily remarked:

> although the Portuguese authorities have now adopted a more active approach towards the dovetailing of state financial policy into the general drive for economic growth (especially the expansion of fixed investment), the principle of a balanced budget has never been questioned; this principle is still sacrosanct, despite the fact that the military expenditure required by the war in Africa represents a substantial part of the State budget, while at the same time the State is increasingly being called upon to prime the pump for the private sector.[42]

The state's obsession with financial stability and its prosecution of the war in Africa left insufficient capital for investment. Rebuffed once again by the state's reluctance fully to commit itself to development through the supply of ample credit, the *grupos economicos* either reinforced their position or purchased a majority of shares in Portuguese banks. Having improved their financial position, they then used the bank's capital to consolidate and increase their holdings within Portugal or to gain leverage with foreign firms and

[40] Portugal, Presidência do Conselho, *Plano Intercalar de Fomento, 1965–67*, i. 20.

[41] See Portugal, Presidência do Conselho, *III Plano de Fomento para 1968–1973*, (1968).

[42] Francisco Cabral, 'Portugal: Towards an Industrial Society', *EFTA Bulletin*, 13: 4 (May 1972), 7.

investors.[43] They formed interlocking directorates and financial alliances so that by the late 1960s, seven large groups controlled the banking system, dominated the Portuguese economy, and increasingly worked independently of the Portuguese state.

Having detailed their colonial holdings, the metropolitan interests of the major groups were the following by 1970. The Companhia União Fabril or CUF which comprised more than 100 businesses, held more than one-tenth of the capital of all existing firms in Portugal, and was the largest of all the groups, with its own bank, Banco Totta e Açores, which had two branches in Africa as of 1973. It had expanded greatly during and after World War II (see Ch. 4) and its principal areas of activity were now chemical products, engineering, petroleum-refining, food products, ship repair, synthetic fibres, maritime transport, paper, and cellulose.

By 1970 the Champalimaud group accounted for two-thirds of sales in the cement industry. With state support, Champalimaud also obtained a monopoly on steel production, the Siderurgia Nacional, in 1954. By 1971 Siderurgia Nacional was responsible for more than two-thirds of the total domestic industrial sales of iron and steel and almost one-half of total sales of base metallurgy. In 1961 the Champalimaud group joined the ranks of the huge conglomerates with the purchase of a majority position in Banco Pinto e Sotto Mayor. Altogether the group possessed twelve industrial firms with interests in cement, paper, steel, banking, and insurance by the early 1970s.

Espírito Santo whose interests centred on the Banco Espírito Santo e Comercial de Lisboa and included banking, insurance, petroleum, agricultural concessions in Africa, paper, cement, beer, and tyres, was already formed in the 1930s but again, traced most of its expansion to the period during and after World War II. It held the largest deposits of all the Portuguese banks, 32 million contos. Its commercial department had 18 million contos. Members of the Espírito Santo family held shares in the bank and were on the Board of Directors of twenty big companies. These twenty firms had a nominal capital of more than 4 million contos, of which 40 per cent was in the colonies.

The Português do Atlantico group was somewhat different from

[43] Ribeiro *et al.*, 'Grande indústria', 966.

the three just mentioned. It was supposed to be less closed and hidebound, more modern in outlook. At its core was the Banco Português do Atlantico founded by Cupertino de Miranda in 1942, but the group was more an association of interests, a combination of diverse commercial and industrial firms which were prominent in the 1940s and linked together in the 1950s. In addition to its substantial colonial interests, within Portugal it had three banks which together held 15 per cent of the capital, 15 per cent of the commercial transactions, and 16 per cent of deposits in the entire country. By 1970 it was one of the largest private shareholders (Espírito Santo, the other) in SACOR, the petroleum refinery and distribution company started in 1939, one-third of whose capital was provided by the state. Banco Português do Atlantico also participated in insurance companies, petroleum-refining, cement, cellulose, beer, glass, synthetic resin, as well as eight other types of industry amounting to a total of 2.5 million contos. During the 1960s, it also became increasingly involved in service industries such as tourism and cinemas.

The Grupo Borges e Irmão which was built around the Banco Borges e Irmão continued to supply credit and maintain investments in its original interests of wine, textiles, and metallurgy and to expand into other fields in the 1950s and 1960s. By the late 1960s, the Grupo Borges e Irmão was involved in forty companies with a nominal capital of 2 million contos. These included banking, insurance, tyres, chemical products, construction materials, textiles, metallurgy, pharmaceuticals, fishing, and tourism. It was a majority shareholder in many of these enterprises, but held minority positions in other companies, too. It had shares also in several national newspapers including *O Primeiro de Janeiro*, *O Jornal do Comércio*, and *O Diário Popular*.

The Grupo Banco Nacional Ultramarino was the only one of the seven groups with a majority of its interests in the colonies, but it also had interests in Portugal. For both areas, the interests included maritime transport, tourism, and insurance. Finally, Grupo Fonsecas e Burnay was only properly consolidated in 1967 by a merger between the two banking groups of Banco Fonsecas, Santos e Viana and Banco Burnay. The holding had international links and engaged in joint ventures with some of the other groups such as CUF, which itself was a shareholder in this group. Grupo

Fonsecas e Burnay also had interests in beer, petroleum-refining, combustibles, as well as colonial interests.[44]

To illustrate the increasing concentration of ownership in industry brought about by the rise of these financial-industrial conglomerates, Table 10.1 shows the progressive decline in the number of employers in Portuguese industry even though the number of employees nearly doubled in the period from 1950 to 1970. In addition, of the 40,051 companies active in Portugal and its islands in 1971, 168 or 0.4 per cent held 53 per cent of the total capital of all the companies:

> Many of these 168 companies are linked together, depend on a central core, in fact form only one group. . . . Beside the statistical fact that there are a small number of large enterprises and a great number of small ones, these links between societies and alliances among groups are the basis of the concentration of capital and production.[45]

Having monopolized many of the most profitable activities within Portugal, these financial-industrial conglomerates now wished to further their own growth through multinational connections and export opportunities, particularly in the EEC. The government raised few objections because it was urgently seeking allies abroad for its African policy. To attract them it needed to feign a more liberal economic policy. Once foreign firms had located in Portugal and the colonies, the government hoped to buy their support for its African objectives. These political imperatives blinded the government to two substantial drawbacks of such a strategy. First, those domestic firms which integrated with Europe and/or formed foreign alliances might lose interest in the colonies and find a colonial war harmful to their international images and objectives. Secondly, integration would disrupt the old balancing act between the modern and traditional sectors, conservative and technocratic supporters of the regime.

The fostering of greater links with multinational capital actually backfired on the Portuguese government. It may have brought temporary foreign support for the regime but it divided the oligopoly, the very group the regime had nursed and protected throughout. Espírito Santo and BNU, with greater domestic invest-

[44] Information on metropolitan investments by conglomerates from Clarence-Smith, *The third Portuguese empire*, ch. 6; and Martins, *Sociedades e grupos*.

[45] Ibid. 16.

TABLE 10.1. *Number of Employers and Employees in Manufacturing, 1950–1970*

Year	Employers	Employees
1950	214,600	624,700
1960	193,200	913,600
1970	76,700	1,019,300

Source: Rafael *et al.*, *Capitalismo*, 13.

ment, greater colonial ties, and fewer European links appear to have been more supportive of the government's colonial policy.[46] The two largest groups, Champalimaud and CUF, on the other hand began to reduce their investment in Angola and Mozambique, to strengthen their links with European multinationals, and increase trade in Europe during the 1960s and early 1970s. But as long as the regime was fighting colonial wars in Africa and repressing opposition at home, Portugal was effectively barred from membership in the European Economic Community. For some of the conglomerates, integration with Europe thus came to mean decolonization in Africa. According to Gervase Clarence-Smith:

> After the death of Manuel de Melo, in 1966, the CUF seems to have shifted its interests towards Europe and Brasil and to have been one of the proponents of peaceful de-colonization.[47]

Champalimaud, at odds with the regime for other reasons, was also calling for decolonization by the early 1970s in order to expand its European interests.[48]

Older industries also came to be divided by the decision to stay in Africa. In textiles, for example, those firms which had modernized and were now exporting abroad took the view of the conglomerates: that integration with Europe was preferable to the costs entailed in continuing links with the colonies. For that reason, many modernized firms within traditional sectors were probably less supportive of the colonial policy. The firms which were most in favour of retaining the colonies were those which had earlier been

[46] Rafael *et al.*, *Capitalismo*, 212–13.
[47] Clarence-Smith, *The third Portuguese empire*, 211.
[48] Ibid. 212.

the subjects of arbitrary and inconsistent government policy: smaller, older factories, unable to move into Europe because of their small size or the poor quality and high prices of their products, were dependent on the colonies for markets and on the government for customs protection.[49] There was no guarantee that a neo-colonial solution was now viable and without protectionism it was unlikely that these industries could have competed successfully in African markets, much less the free trading environment of the European Community. But as long as Portugal maintained enclaves where Portuguese products were bought and sold under conditions highly favourable to the metropole, the European Community treated its application unfavourably. When the Portuguese government signed a negotiating mandate with the European Community in 1971 (faced with the near-certainty of UK membership) the colonies were not included—a clear message that the Community did not share the regime's imperial preferences. Because of these market and capital considerations, all industrial groups then appear to have been divided by the wars and this weakened support for the government.

TROUBLES AT HOME

The difficulties that the colonial wars presented to integration with Europe were only one issue, other domestic developments added to the political pressure on the regime. Among the more important factors can be cited the crisis which continued in many of the older industries, the cleavage in the government between technocrats and conservatives, the low standard of living, escalating emigration, rising inflation due to the wars and foreign investment, and political repression.

In the 1950s, many of the older sectors had complained about the government's failure to support modernization efforts. Although a small, select group of firms had modernized in several of the important traditional sectors such as wine, textiles, cork, and food-processing, the largest number had not. Despite the promises to speed modernization and reorganization along the lines set forth in the Second Development Plan and the creation of new organiza-

[49] Rafael *et al.*, *Capitalismo*, 100; and Pereira, *Indústria, ideologia*, 24.

tions such as corporations and *grémios* to improve the representation of these sectors, these efforts did little to alleviate their general malaise, for the means were unsuited to the ends. In his study of the industrial élite, Makler found that 'some industrialists [in the corporatized sectors] expressed strong skepticism about the corporative system in general' and felt that *grémios* did not adequately represent their interests.[50] Many of these industries were still in crisis in the mid-1960s, unable to sell their products at profit or to find adequate markets. While the Third Development Plan of 1968–73 envisioned the fusion and concentration of smaller factories in the traditional sectors, firms had neither the interest nor the capital to merge with other firms who were equally debt-ridden and technologically backward.[51]

To those who favoured modernization, fusion, and integration with Europe, were joined a new breed of technocrats working for the government. The complaints of the late 1950s had struck a chord in this group and during the 1960s, they began to express their discontent with some of the objectives of government policies. They believed that Portugal should join the twentieth century through the modernization and expansion of industry, improved educational training, and the implementation of high-growth policies. To this group, whose views come through clearly in the Third Development Plan, the colonial wars epitomized the backward image that Portugal had in Western Europe. So the colonial wars, though not directly related to the modernization issue, deepened the growing rift within the regime between the modernizers and the conservatives.[52]

The regime could probably support a little grumbling here and there, after all industrialists had been complaining about markets, prices, competition, and government intervention since 1946; a more serious threat arose in the form of emigration. Although industrial production expanded by almost 7 per cent per year during the 1960s and early 1970s and Portugal's GDP increased from $US275 in 1960 to $US1,271 by 1973, the growth experi-

[50] Makler, 'The Portuguese Industrial Elite', in Graham and Makler (eds.), *Contemporary Portugal*, 146.

[51] See *III Plano*, vol. i.

[52] Rafael *et al.*, *Capitalismo*, 99–100. This rift was also paralleled in the military by the late 1960s though it originated in professional as well as political grievances, see Bandeira, 'The Portuguese Armed Forces'.

enced by Portugal's industrial sector had not benefited workers proportionately. Low salaries had enabled the modern export-oriented industries to make large profits from sales in Europe and America, formed part of the attraction for foreign investment in Portugal, and were partly responsible for the preservation of many unproductive industrial units, as in the textile industry. Miriam Halpern Pereira observed that:

> In reality, the policy of low salaries made possible the simultaneous survival of backward industry and the development of export industries, based in the pre-existing industrial structure and in the modern sector where foreign or foreign-national capital are very numerous.[53]

Because of the appallingly low salaries in agriculture and industry, the threat of three-year military service in the colonies, and increased possibilities for employment in France and Germany, emigration from Portugal (which had stabilized at around 34,293 emigrants per year during the 1950s) increased dramatically from 1963. According to official statistics, between 1961 and 1973, an average of 66,546 people per year emigrated from Portugal. This did not include clandestine emigration which averaged 39,345 emigrants per year, climbing precipitously by the end of the decade to peak at 106,907 illegal emigrants in 1970.[54] Altogether it is estimated that approximately 10 per cent of Portugal's total population emigrated between 1961 and 1974.

Once again emigration coincided with pressing financial needs of the government and it was loathe to stem the tide, though it is unlikely that the government foresaw the scale of emigration. Up until 1957, the government had used transactions with the colonies to cover deficits in their own balance of trade. But that year, deficits occurred in spite of transactions and the government looked again at rising emigrants' remittances to cover the deficit. A series of events then made increasing remittances not only fortuitous, but also imperative for the regime's financial strategy. First, the drive to modernization and the demands made by all industrial sectors that the state contribute support required that the regime resort to capital expenditure. Secondly, after 1961, the government needed to finance the colonial wars. Between 1961 and 1974, military expenditure averaged around 40 per cent of the government's

[53] Pereira, *Assimetrias*, 44.

[54] Leeds, 'Labor Export, Development', 273.

budget[55] and was three times the amount spent on development in 1972–3. Indeed, military expenses probably account for budget deficits in 1972 and 1973 of 1.6 million contos and 3.6 million contos, respectively.[56] Thirdly, to satisfy colonial and metropolitan commercial interests and thereby to garner support for its policies there, the government had to liberalize financial relations and relax its controls on imports into the colonies. The escudo zone which had long been trumpeted as Portugal's own special free trade zone had to become a reality in 1963. In that year, not only was it no longer necessary for the colonies to pay in advance for goods imported from Portugal, but also import controls for goods coming from abroad were relaxed in order to speed up colonial development projects. In a throwback to the delayed payments problems of the Republican regime, the colonies had fallen behind on payments of goods imported from Portugal within a year, crippling many traditional sectors such as textiles and undermining Portugal's trade balance. All of these factors drained the state's resources and made it increasingly unlikely that the colonial foreign-exchange funds could make up the difference, even after the reimposition of exchange controls in 1971. Therefore, the government resorted to a trusted means of covering a negative balance of payments, emigrants' remittances. Excluding colonial transactions with foreign countries, Portugal's negative balance of trade of 4,702,000 contos in 1964 became a positive balance of payments of 1,366,000 contos. In 1973 a trade deficit of 18,349,000 contos became a payments surplus of 8,403,000 contos with the addition of remittances.[57] In the escudo zone the share of remittances in covering the deficit went from 47.2 per cent in 1964–8 to 76.4 per cent in 1969–73.[58] Having become dependent on emigration to preserve its own financial health, the government was unlikely to prevent it, although it did take measures to thwart it until the late 1960s.[59]

Although emigrants' remittances resulted in increased consumption in Portugal as well as foreign exchange for the government (as in the Republican period), Portuguese businesses could not meet

[55] Rona M. Fields, *The Portuguese Revolution and the Armed Forces Movement* (New York: Praeger, 1976), 56.

[56] Robinson, *Contemporary Portugal*, 176.

[57] Rocha, 'Colónias e exportação', 1063.

[58] Ibid. 1068.

[59] See Leeds, 'Labor Export, Development', ch. 6 on emigration policy in the 1960s.

the added demand because they had lost their workers. In particular, agricultural enterprises and the less efficient, small, labour-intensive firms were hurt most by emigration. Although fixed prices for many agricultural products had initially been higher than prices for other products, these prices remained unaltered during the 1950s and were now lower than other prices.[60] Coupled with the fact that industry was increasingly orientated towards exports and not towards the internal market, inadequate agricultural prices and lack of reform discouraged or prevented investment in this sector. In 1968, for example, only 3.3 per cent of agricultural businesses invested in new crops or introduced new methods of cultivation even though 30 per cent of the active population was still on the land.[61] Additionally, powerful landowners themselves were resistant to change:

> The regime has been unable or unwilling to transform agrarian structures or induce new patterns of entrepreneurial behaviour in the primary sector, especially in the southern *latifundia*.[62]

Agricultural exports declined in importance between 1953 and 1965.[63] The stagnation of agriculture (and colonial wars) encouraged widescale emigration by agricultural workers and peasants from the 1960s.

As people left the land, crops remained unharvested and agricultural prices rose due to the drop in supply.[64] For those who had not emigrated the higher prices resulted in an added burden upon an already difficult situation. Inflation set in due to price increases and military expenses resulting in general economic discontent and further emigration. Wages were then raised in response to the inflation and to stem emigration. This again affected the less efficient firms because they had depended on low wages (and therefore low food prices) to survive and did not have sufficient profit levels to cover the increases.[65]

[60] Armando Sevinate Pinto *et al.*, *A agricultura portuguesa no período 1950–1980* (Lisbon: Imprensa Nacional, 1984), 17.

[61] Pereira, *Assimetrias*, 38–9, for rates of investment; and Martins, 'Portugal', in Archer and Giner (eds.), *Contemporary Europe*, 72, for sectoral employment of the labour force.

[62] Martins in Woolf (ed.), *European Fascism*, 330.

[63] Andrew Wilson Green, 'Portugal and the African Territories: Economic Implications', in Abshire and Samuels (eds.), *Portuguese Africa*, 346.

[64] Pinto *et al.*, *A agricultura*, 28.

[65] Pereira, *Assimetrias*, 51–2.

Criticisms from labour, small and medium businesses, and the agricultural sector swelled the persistent criticisms launched by the different opposition groups in the country. Together with these economic grievances, came attacks on the repression of the *Estado Novo*. Though opposition groups abstained from both the 1961 and 1965 elections they continued to criticize the lack of civil liberties in Portugal. The student movement became particularly well-organized and vociferous during the 1960s and early 1970s.[66] Even Monarchists and Catholics, former supporters, started criticizing the regime for the lack of freedom and other political liberties such as freedom of the press and freedom to organize.[67] Most importantly, by the late 1960s, almost all opposition groups accepted the right to self-determination in the African colonies. Caetano's constitutional revision in 1971 which provided greater political and economic autonomy for the colonies and resulted in the election of many Africans to newly created legislative bodies did little to dampen the opposition's demands or to weaken revolutionary movements in Portugal's African colonies.

Self-determination in the colonies became linked with the defeat of authoritarianism at home. With questions of freedom now on the agenda because of the war, it would have been politically difficult for the authoritarian government to grant independence in the colonies without granting political liberties in the metropole, and difficult to prosecute the colonial wars if liberalization were allowed at home. In the economic arena, the regime only just managed to appease the major economic groups in this period; Salazar's long experience in balancing the different forces at play helped keep the regime afloat during the 1960s—postponing intractable problems for his successor. After unsuccessfully attempting to liberalize further upon coming to power in 1968, Caetano marked his retreat to old-style authoritarianism and interventionism by replacing technocrats with hard-liners in key government economic positions in 1972.[68] Unable to liberalize economically without granting political freedoms, Caetano resorted to previous tactics: control, censorship, regulation, repression.

[66] Robinson, *Contemporary Portugal*, 79; and Martins, 'Opposition in Portugal', 261, Caetano himself resigned as rector of Lisbon University in support of a student strike in 1962.

[67] Robinson, *Contemporary Portugal*, 79–80.

[68] Rafael *et al.*, *Capitalismo*, 142.

The colonial wars thus encouraged economic growth but paradoxically, exacerbated what was already a difficult domestic situation by the early 1960s. Grievances about the pace of development, the lack of markets, government favouritism, the stagnation in the older industries, and the lack of political freedom abounded. The colonial wars provided a temporary distraction but once the knee-jerk nationalism subsided, the old problems resurfaced, worsened by the wars which fed inflation and emigration and provided a major stumbling-block to European integration and the formation of international alliances. Salazar skilfully managed the contradictions until 1968, when, after a disabling fall from a deck-chair, he was replaced by Marcello Caetano. Although Caetano tried to achieve that mixture of liberalization and continuity which Salazar had advocated, the contradiction between economic flexibility and political rigidity deepened. European integration and the defeat of authoritarianism became synonymous with the withdrawal from Africa.

11

A Divided Industry

In the hierarchy of government priorities in the 1960s and early 1970s, the needs of the textile industry ranked below the economic development of modern or non-traditional industries and later, the quest for domestic and colonial political stability. During the period, the textile industry continued to suffer from insufficient markets, cutthroat competition, and governmental indifference to the industry's production and export problems. The commencement of the colonial wars and high emigration during the 1960s compounded the industry's problems. By 1974 the textile industry had become sharply divided both structurally and commercially and, as a whole, had lost status to growing financial-industrial conglomerates.

Although the entire industry experienced a profound crisis during most of the period we are discussing, sectors within the industry not only responded differently to the crisis, but were treated unequally by the government. Government preferences for larger, technologically advanced firms led to expansion and modernization of these during the 1960s and 1970s. Moreover, government-supported schemes such as the Stabilization Fund in 1959, Portugal's participation in the Stockholm Convention, and its entry into the European Free Trade Association in 1961 further aided this sector. Government favouritism was inconsistent, however, and many of its other priorities contradicted its intentions to modernize the industry and increase textile exports. Such actions affected the potential of larger, modern firms to take full advantage of new machinery and bigger markets.

For smaller and older firms in the industry, Industrial Regulation laws, domestic competition, insufficient credit, limited export opportunities, and emigration repeatedly jeopardized the survival of many and damaged further an already ailing sector of the textile industry. Due to the different capabilities of the two sectors and variable governmental treatment, the industry split between large, modern firms producing textiles for export abroad and small, antiquated ones trading in the domestic and colonial markets—replicating the bifurcation between modern and traditional industry in the society as a whole. In this chapter, we explain the origin of division by looking at the factors which have influenced the industry throughout its history: prices, the cotton supply, structure, legislation, and markets. Finally, we examine the contradictions in government policy in order to understand the prolongation of the crisis throughout the 1960s and early 1970s.

1958 AND AFTER: OLD PRIORITIES, NEW HORIZONS

While central government and Lisbon voters focused on Delgado's presidential campaign during the 1958 elections, in the North, the troubles of the textile industry emerged as the central topic of debate. That year, João Dias Rosas, the former President of the Cotton Regulatory Commission, declared himself a candidate for deputy to the National Assembly from the District of Braga. Braga was one of those districts with a preponderance of small firms and had been particularly hard hit by the crisis. During the campaign for deputy, Dias Rosas tried to play down the problems of the industry by asserting that the crisis had ended in 1957. He also commented that many companies had modernized and had expanded their output and those who were unable to adapt were badly equipped, possessed poor working conditions, and had incompetent management.[1]

Predictably, Dias Rosas was elected but one of his critics produced embarrassing evidence of government neglect of the textile industry. Claiming to represent the view of textile industrialists, Armando Carneiro, the self-appointed Director of the Gabinete de Estudos de Divulgação Económica e Social (GEDES) and a genuine corporatist, produced a study of the industry which contradicted the views of Dias Rosas. Carneiro found that the crisis was not limited to a few incompetently managed, badly equipped firms as Dias Rosas had alleged. On the contrary:

> Except for a half dozen mill owners . . . everyone—big, medium, and small—feels the crisis. And although they do not sound as if they have

[1] Carneiro, *A crise*, Part I, 22.

given up, I feel that these people are very discouraged, and that the disdainful and meddling sarcasm of their opponents increases day by day.[2]

For this predicament, Carneiro blamed the 1946 law because 'during the last ten years, there have been those who could neither legally replace machinery nor achieve equilibrium'.[3] He publicly exposed the government's favouritism towards large factories and singled out the tenacity of small manufacturers:

In passing, I simply want to call attention to the merits of the indefatigable small manufacturer, whom the technocrats unanimously condemn, but who carries on against all odds, working and resisting, naturally exhausted and impoverished, but fighting, always fighting against adversity.[4]

Carneiro called upon the government to conduct a study of, and to create a development fund for, the textile industry. Furthermore, he requested that the government locate external markets through diplomatic action, lower taxes, and increase the standard of living in Portugal and the colonies so that consumers would purchase more textiles.[5]

The government banned Carneiro's study group after the publication of his pamphlet on the textile industry. The main bone of contention seems to have been his suggestion that the textile industry form its own *grémio*. As we noted, when other industries were joining *grémios* in the 1930s, the textile industry did not seem eager, nor did the government insist that a *grémio* be formed. Instead, textile industrialists belonged to the AIp but as Carneiro charged, membership was expensive and exclusive. A *grémio* of textile industrialists would have a wider membership, greater authority, and facilitate relations with official entities, he claimed. Shortly after this suggestion, the government shut Carneiro's office and suppressed the results of his study.[6] An old-fashioned corporatist had embarrassed a government which claimed to espouse the ideals of corporatism but had not yet organized a *grémio* for its largest industry.

In spite of Carneiro's treatment, however, 1958 proved to be something of a watershed for the textile industry. The year marked

[2] Ibid. 62. [3] Ibid. 79. [4] Ibid. 73.
[5] Ibid. 88–97. [6] Ibid. 117–39.

the beginning of more resolute (but not necessarily favourable or satisfactory) government action concerning technical difficulties, cotton prices, and exports. That year, the government created the Corporation for Industry (Corporação da Indústria) with a separate section for textiles. It also appointed a Committee of Reorganization to study the problems of the textile industry and in 1959, it established the National Institute of Industrial Research (Instituto Nacional de Investigação Industrial), which also had a separate textile department. The government also sponsored and approved the formation of a *grèmio* for the textile industry in 1963—reflecting a feeble attempt by conservative government officials to resurrect corporatism.

In addition to general development plans, incentives to foreign investment, and more liberal trade and investment policies during the 1960s, the government sought and secured external markets on favourable terms to further aid the export-oriented sector of the textile industry. In 1959 Portugal signed the Stockholm Convention, a precursor to its membership of EFTA in 1961. Membership brought most favoured nation status to Portuguese textiles in markets of member EFTA countries. Moreover, the government also signed agreements with the American and Brazilian governments to export textiles to those markets. In 1969, after Caetano became Prime Minister, the government renewed a pledge to procure export markets for the industry and made good on that pledge in the 1971 trade agreement with the European Economic Community. The 1971 trade agreement followed by the Special Relations Agreement of 1 January 1973 established generous ceilings for Portuguese textile exports to member countries of the European Community.[7] These measures testify to an export-oriented approach by the government throughout the 1960s and early 1970s and did produce dramatic increases in exports.

What the government would not do in 1958 was alter its colonial cotton pricing policy. While this should have come as a relief to small industries who had benefited from the low prices through the operation of the quota system, the situation had altered radically by 1958. From 1951 the international price of cotton had been dropping. In 1956 it had fallen slightly below Portuguese prices but

[7] Francis Nicholson and Roger East, *From the Six to the Twelve: The Enlargement of the European Communities* (Harlow: Longman, 1987), 242–4.

allowing for customs duties and import taxes on foreign cotton, this difference was probably eradicated. By 1958, however, the international price had fallen 1$48 below the colonial cotton price to 16$82 (Table A7). Prices of colonial cotton were now higher than foreign cotton and Portuguese textiles risked losing their already meagre share of the international textiles market. In 1958 the Companhia da Fábrica de Fiação de Tomar reported:

> Not only the decrease in the volume of exports due to the low price of foreign cotton in relation to colonial cotton but also the decrease of purchases by our overseas province of Angola, has resulted in the introduction in the internal market of an increased number of manufactured products and consequently a drop in prices which an over-supply causes.[8]

The Fábrica de Fiação e Tecidos de Santo Tirso reiterated exactly the same thing in its 1958 report, mentioning the drop in exports and the consequent decline in domestic prices due to the drop in the international price relative to the colonial price of cotton. This company, though, remarked that it had modernized and was in a better position than some factories to compete domestically and abroad.

Despite the alarm sounded by these and other mills, stability and the encouragement of colonial production took precedence over the benefits of lower prices to the industry. Since the government would not alter the colonial pricing policy, the choice was either to do nothing, risk a drop in exports and a rise in textile imports, or provide subsidies. In 1958, under pressure from the CRCAR and textile manufacturers, it agreed to subsidize textile exports abroad.[9] A year later the establishment of a Stabilization Fund on 9 July 1959 proposed a neat solution which would keep the industry tied to colonial cotton while not prejudicing it in export markets.

In the preamble to the decree, the government recognized the difficulties of the industry and the need for an emergency solution for its structural deficiencies. To aid the industry, taxes and import duties on colonial cotton were lowered and export duties on goods made with colonial cotton were dropped. Shipping charges were reduced by 10 per cent for cotton and cotton textiles. Secondly, the government agreed to pay importers of colonial cotton the difference between colonial and international prices, but only if the

[8] Tomar, *Relatório*, 1958, no page.
[9] Fonseca, 'Intervencionismo', 132.

cotton was destined for manufactures for export. Similarly, the government dropped customs duties on imports of foreign cotton if the cotton was used in goods made for export. In effect then, the internal market benefited from tax and duty reductions, and the regime subsidized importers and exporters with connections to foreign markets. The colonial cotton price was thus maintained at its old level in order to satisfy 'other national interests', the government claimed, and the taxpayer would have to pay the difference.[10]

Since the quota system was still in operation, presumably in the years when the colonial cotton supply was insufficient, larger firms would benefit in the domestic market from imports of cheaper foreign cotton just as prior to 1958 they had suffered from imports of more expensive foreign cotton. In export markets, larger factories now faced the prospect of competing on equal terms with factories in other countries. To make sure that larger mills were not disadvantaged by poorer quality colonial cotton, the government eased the restrictions on imports of foreign cotton. This is evident from the increase in foreign cotton imports and suggests that the government had accepted the complaints about the low quality of colonial cotton from larger textile firms. To compete effectively abroad, higher quality cotton was needed.

The decisions taken after 1958 mirrored many of the general developments occurring in the Portuguese economy. Just as the government aided the financial-industrial conglomerates so also did it aid the advanced sector of the textile industry through credit schemes and trade agreements. As with its assistance to financial-industrial conglomerates, its policies were not comprehensive and were fraught with contradictions, reflecting the divisions within the government. Alongside technocratic enterprises and liberal economic policies, many of its initiatives such as the creation of a *grémio* and the creation of a Corporation for Industry indicate the persistence of a corporative agenda. The regime also clung to its old priorities: stable prices, links with the colonies, the quota system, and Industrial Regulation. These contrary priorities were pursued in textiles through a division of markets: the export market for larger firms, the domestic and colonial markets mainly for smaller, older firms. The government also divided cotton prices and supplies. Colonial cotton prices were maintained at the old level in

[10] *Diário do Governo*, Série I, 9 July 1959, D-L 42,375.

the domestic market but equilibrated to international prices in markets abroad; and foreign imports were eased to ensure supply and improve the quality of exports.

In opting for a divided policy for textiles, the regime was seeking a number of political and economic objectives. Rather than try to eliminate smaller, older mills through direct administrative measures, the government adopted an indirect approach through markets and prices thereby hoping to preserve economic and political stability or, at least, to avoid responsibility for any crisis. Secondly, in accordance with the general, but confused orientation of policy, the government sought both to pursue integration with Europe and to achieve unity with the colonies. In textiles, approximately one dozen large, advanced firms would help to cement commercial integration with Europe and hundreds of smaller, antiquated firms would maintain ties with the colonies by buying cotton and selling finished textiles. One rule existed for the large firms in the foreign export sector and another for small producers in the domestic and colonial markets: the former was first helped to modernize and later, to expand into markets abroad; the latter was preserved in aspic, at the expense of metropolitan and colonial consumers.

THE INDUSTRY DIVIDES

Years after these measures were passed, companies were still complaining about the crisis in the industry. Statement after statement depicted companies struggling to survive. According to the Companhia da Fábrica de Fiação de Tomar, cutthroat competition and price drops continued to prevail and businesses had gone into debt trying to compete in these conditions:

> This crisis in the textile industry was produced by the inordinate expansion of productive equipment which the financial facilities then present invited and that a reprehensible or badly interpreted Industrial Regulation did not prevent.[11]

Frustrated, the company declared in 1968, 'we do not accept the general belief that textiles is no longer the greatest national industry. Which is the greatest then?!'[12]

[11] Tomar, *Relatório*, 1967, no page.
[12] Ibid. 1968, no page.

In response to the frustration expressed by the industry, the government appointed a commission (Comissão de Estudo da Indústria Têxtil) in 1967 to examine and solve the following problems in the textile industry: the supply of cotton to the industry, the system of finance for replacing machinery in those factories of sufficient size, the industrial structure which would be most appropriate for future competition in both the domestic or external markets, and the technical and professional aid appropriate to industrial reorganization.[13] Cotton supply, markets, and the structure of the industry: these were the same problems which had plagued the industry in the 1930s, 1940s, and 1950s. What changes had the 1960s brought?

The commission found that the measures taken after 1958 had only masked the crisis, they had not addressed its origins and moreover, worked imperfectly in practice. Though parts of the industry were now affected differently by the crisis, old problems lingered and new ones arose. The government's cotton policy as well as its preoccupation with the colonial wars also produced contradictions in the relationship between the government and the textile industry during the 1960s. Below we discuss some of the major elements of the industry's persistent crisis as well as attempts by the Caetano Government to solve them prior to 1974.

STRUCTURE AND FINANCE OF MILLS

Structural and financial difficulties remained one of the most vexing problems for the industry during the 1960s and 1970s. Many companies continued to refer to over-production in the internal market and to cutthroat competition as well as to debt and financial collapse. What occurred can be ascertained by an examination of the different responses by companies to the creation of the Stabilization Fund, the entry of foreign investment, and the conclusion of trade agreements with other countries. First, many firms took advantage of the opportunities on offer in the late 1950s and early 1960s and modernized or expanded. It was during this decade that large firms such as Manuel Gonçalves, Riopele, Somelos, and Coelima began to appear, expanding vertically and horizontally

[13] Comissão, *Relatório*.

and forming links with foreign firms. These firms were orientated towards the high-quality end of the market and produced mainly for export. Some of them also expanded into the manufacture of synthetics at this time.

Total investment between 1959 and 1966 equalled 3.5 million contos, 76 per cent of which went to the purchase of new machinery.[14] A comparison of 1958 to 1966 industrial figures confirms that investment took place. The total number of factories only increased by eight between 1958 and 1966; yet the total number of spindles grew by about 200,000 and looms by about 5,000. Since 1958, 9,140 automatic looms had been installed, so replacement of old by new machinery had occurred.[15] Growth took place mainly in the larger plants; factories possessing between 20,001 and 40,000 spindles increased from 9 to 16 between 1958 and 1966 and those with over 100 looms went from 71 in 1958 to 97 in 1966 (see Tables A14 and A15). Some of the increase could be attributed to factories which increased their machinery and therefore shifted upwards to another category. These categories also accounted for most of the increase in machinery in the ten-year period.

Secondly, many firms neither expanded nor modernized and many may have collapsed. Not counting cottage industries estimated at approximately 300 with 11,000 looms, the 1968 commission found that among the firms covered by Industrial Regulation, many were still small: 75 per cent of weaving plants had less than 100 looms; 25 per cent of the total number of looms were installed in these plants—whose production constituted a sizeable threat to price stability according to the commission.[16] This may be true but survival was tenuous given their financial state. Ignoring the plight of small and old firms, the government suggested that several firms had simply taken advantage of the Stabilization Fund and had not used its benefits to buy new equipment. These were the reasons cited by the government for its yearly extension of the Stabilization Fund up to 1965.[17]

As figures show, whatever their size, many factories simply did

[14] *Diário do Governo*, Série I, 12 Feb. 1969, Despacho Ministerial.

[15] Comissão, *Relatório*, 17.

[16] Ibid.

[17] *Diário do Governo*, Série I, see D-L 43,857 (12 Aug. 1961); 44,605 (27 Sept. 1962); 45,285 (2 Oct. 1963); and Portaria 20,379 (19 Feb. 1964).

not modernize. The commission found that out of a total of 37,250 looms, nearly half were more than twenty years old.[18] Weakened by the crisis of the 1950s, it is probable that many firms did not have the financial capabilities to embark on a modernization programme. In company reports, the factories of Ermezinde and Crestuma both lament the obsolete state of their equipment and their weak financial situation. Moreover, the death of the Companhia Fiação e Tecidos do Porto, one of Portugal's oldest factories depicted the worst nightmare of many textile manufacturers. Porto attempted to substitute its old machinery in the early 1960s, but poor sales abroad and at home from 1964 to 1969 prevented the company from modernizing further. By 1968 the company was calling for 'urgent measures to be adopted in order to escape from this critical situation' but these measures never materialized.[19] A fire in Porto's weaving section in 1970 prompted the management to close the factory as it lacked resources to replace the machinery destroyed in the blaze. By 1973 most of the machinery had been sold and the buildings purchased by Pão de Açucar, a large supermarket chain.[20]

However, as several firms in their reports referred to the purchase of new equipment and the reorganization of plants to better compete in internal and external markets, a third response must be considered. According to one textile manufacturer, Lino Lima:

> Legal requirements of the Stabilization Fund along with the possibility of placing their textiles in export markets led some industrialists to invest in new machinery even above their means. Some industrialists feared they would be shut by the government if they did not modernize and secured loans for that purpose. Since many did not have the necessary capital to obtain long term loans, (which were scarce, at any rate), they applied for short term loans at high interest. In this they were encouraged by the banks who used to say; 'March ahead with your know-how and my money.'[21]

Relying on credit, industrialists had rushed to modernize in order to take advantage of new trading agreements, an explanation reiterated in a government dispatch on the industry issued in 1969. According to the government, of the 3.5 million contos invested in

[18] Comissão, *Relatório*, 17.
[19] Porto, *Relatório*, 1968, no page.
[20] Ibid. 1970–3.
[21] Lima, 'A crise', 14.

the period from 1959 to 1966, 2 million contos was short-term credit, the majority of which was borrowed by firms in excess of their available operating capital.[22] When the banks started to call in the loans, industrialists then bought raw materials on medium-term credit from cotton importers, thus involving intermediaries as well. Many of these companies, mostly small and medium-sized, then went into liquidation.[23] The Borges e Irmão group (referred to earlier as having financed textile companies) was listed as a creditor of several firms which collapsed in 1966–7.[24] By 1968 then, two out of the three responses had resulted in debility or collapse and the situation had reached catastrophic proportions. According to Lima, the crisis could only be solved by 'conscious and determined action by industrialists, banks, and the Government who, through a series of careless actions, are cumulatively responsible for the crisis'.[25] The government admitted that there was a crisis in 1967 but blamed it on the industry; the government's determined action was to appoint the study commission.

In 1969, following a report by the commission to the Caetano Government, a number of rules were then established for those companies considering a recourse to credit to finance expansion and modernization. Short-term credit would be converted to medium- and long-term credit if the company's capital and reserves covered at least 30 per cent of the investment; if loan repayment time did not exceed 80 per cent of the depreciation period of new equipment; and if the company possessed the technological, administrative, and commercial conditions to achieve a satisfactory return on its investment. The government offered tax benefits and exemptions to those firms who merged with or incorporated other firms.[26] Even if one of the signers of the dispatch had not been João Dias Rosas, a former President of the CRCAR, now the Minister of Finance and an advocate of technocratic solutions to industry's ills, government bias towards large, modern firms was very clear. Government preferences had not changed: incentives designed to benefit larger firms had merely supplanted regulations implemented to control small mills.

22 *Diário do Governo*, Série I, 12 Feb. 1969, Despacho Ministerial.
23 Lima, 'A crise', 15.
24 Martins, *Sociedades e grupos*, 51.
25 Lima, 'A crise', 12.
26 *Diário de Governo*, Série I, 12 Feb. 1969, Despacho Ministerial.

By the 1970s, like so many times previously, these incentives had not produced significant improvements in spite of government propaganda devoted to aiding the industry in the 1960s. A drop in the total number of factories from 436 in 1966 to 396 by 1974 plus increases in the total number of spindles and looms may indicate fusion, but the evidence equally suggests that the industry remained in crisis. Company reports continued to speak of financial difficulties. Moreover, only 92 firms in 1974 employed over 200 workers compared with 84 in 1966; and 101 mills (not including cottage industries) employing less than 20 people still officially operated compared with 97 in 1966. Neither of these figures suggest that the 1969 legislation changed significantly the state of the industry.[27]

MARKETS FOR TEXTILES

For the large, modern textile mills geared principally towards foreign markets, textile industrialists vacillated between optimism and exasperation at government efforts to secure these markets and at the response from abroad during the 1960s and early 1970s. Initially, as a result of Portugal's signing of the Stockholm Convention textile exports abroad more than trebled from 6,692 tons in 1959 to 19,641 in 1960 surpassing, for the first time, exports to the colonies which had remained more or less stable at 10,000 tons throughout the 1950s. Between 1960 and 1970 as a result of adhesion to EFTA plus government subsidies provided by the Stabilization Fund (after 1965, the Export Development Fund), textile exports abroad increased by 36 per cent.[28] By 1970 total textile exports abroad equalled 30,668 tons—four times the amount of exports to the colonies. Manufacturers revealed the extent of their dependence on these markets by the magnitude of their praise:

[27] Portugal, INE, *Estatística Industrial*, 1966–74. In 1970 the INE stopped giving the total number of spindles and looms in the industry, judging instead the size of factories by the number of workers employed. Grémio Nacional dos Industriais Têxteis, *Informação*, 3 (Jan./Feb. 1974), 3, has figures on spindles for 1972 and looms for 1970 and see Tables A8–A10.

[28] See also Pedro Alvares and Carlos Roma Fernandes, *Portugal e o Mercado Comum*, i, 2nd edn. (Lisbon: ed. Portico, 1980).

The policy of export development for the world market intelligently fostered by the Government of the Nation, principally through the maintenance of an indispensable export subsidy, brings an appreciable increase of sales abroad to this important sector of national industry. This increase in sales abroad, while still slow, is becoming more and more necessary either as a generator of foreign exchange, or as a condition of survival in the future fusion of Western markets. Were these measures not maintained, we would face enormous difficulties.[29]

This dependence was graphically illustrated several times during the 1960s and early 1970s. When intermittent protectionist policies or cancelled orders in foreign markets contributed to reduced demand, exports shrank. Large firms then dumped unwanted textiles on to the highly competitive domestic market, as occurred, for example, when exports abroad went from 44,013 tons in 1972 to 22,861 tons in 1973—the year Great Britain joined the EEC.

For those firms which were small and old, the metropolitan and colonial markets offered only periodic relief. The domestic market continued to be oversupplied and highly competitive. Consumer buying power had not increased markedly since the 1960s due to low wages. In the colonial market, the initial outbreak of war in 1960 affected the industry both directly and indirectly. Fighting in the colonies initially disrupted exports from the metropole to there and this greatly worried some textile industrialists. As violent clashes took place in Angola, one company remarked:

Unhappily the very serious events that have affected our overseas provinces, notably Angola, have resulted in nearly total suspension of sales and upset the market completely, profoundly affecting the results of the first half-year.[30]

Nehru's takeover of Portuguese India also provoked a reaction as one company report attested:

The restriction of transfers and consequently the drop in trade due to the disturbances in our province of Angola and the disgraceful occupation of Portuguese India . . . resulted in a less favourable year for the Portuguese textile industry.[31]

[29] Rio Vizela, *Relatório*, 1961, no page.

[30] Fábrica de Fiação e Tecidos de Ermezinde, *Relatório e contas de direcção e parecer do conselho fiscal*, 1961, no page (hereinafter Ermezinde, *Relatório*).

[31] Rio Vizela, *Relatório*, 1961, no page.

By 1962, however, exports to the colonies returned to their old level. Due to increased needs occasioned by the war, textile exports to the colonies swelled in the early 1960s. In 1964 and 1965 Portugal exported more than 15,000 tons of textiles per year to the colonies, almost one-third more than its average for the 1950s. No doubt this particularly pleased many small and medium industrialists who were wholly dependent on the colonies for sales. Thus support for the colonial policy may have come from the very sector which the regime had earlier tried to eliminate.

Growth unfortunately stalled after 1965. As a consequence of the regime's efforts to create a Portuguese common market in the colonies and to pacify settler discontent, the payments system between the colonies and the metropole was liberalized in 1963 and textile industries in the colonies were allowed to expand their capacity from 1965.[32] Regarding interterritorial payments, the colonies were no longer required to pay for their imports in advance of purchase, signalling a return to a method of payment which had prevailed during the Republican period. The results of the change also mimicked the Republican experience. Just as had occurred prior to the regulation of interterritorial payments in the 1930s, liberalization produced delays in financial transfers for goods purchased from the metropole.[33] Since the textile industry still sent 30 to 50 per cent of its exports to the colonies, these delays, combined with credit problems, intensified the industry's crisis as companies like Fafe, Rio Ave, Jacinto, and Ermezinde repeatedly attest. In 1970 Fafe reported that delays from the colonies had put the industry in a 'grave situation' and in 1971, the Fabrica de Fiação e Tecidos de Ermezinde argued that its financial position was at risk owing to its dependence on clients who had 'tens of thousands of *contos* tied up in the colonies'.[34] By 1972, in an about-face, the government had returned to the previous system but as Rocha remarks, 'the new system had a short life'.[35] After 1974 any industrialist who wanted to trade with newly independent Mozambique and Angola did so at his own risk.

Additionally, the growth of textile factories in the colonies prior to 1974 directly competed with metropolitan industries resulting in

[32] See Sociedade Algodoeira de Fomento Colonial, *Relatório*, 1965–6.
[33] Rocha, 'Colónias e exportação', 1061.
[34] Fafe, *Relatório*, 1970; and Ermezinde, *Relatório*, 1971.
[35] Rocha, 'Colónias e exportação', 1062.

a reduction of metropolitan exports and the dumping of textiles back on to the internal market. For those large textile mills who had invested in these factories in the 1940s and had now shifted their own exports to Europe, the beginning of local production evoked mild interest, since the return on the investment was never very great. But for those metropolitan mills which were heavily dependent on the colonies, the growth of local production introduced an additional challenge to an already ailing industry. The gradual drop in exports to the colonies after 1965 indicates that local industries were supplying the domestic markets of Angola and Mozambique. While some of the local textile sales could be accounted for by an increase in textile imports from other countries, cotton producers in both Mozambique and Angola were supplying local factories with sizeable amounts of cotton by this date. Moreover, after sluggish growth in the 1950s, factories owned by the Sociedade Algodoeira de Fomento Colonial in both Angola and Mozambique expanded their capacity after 1965 and reported gains in sales brought on by the war. By 1971, 15 per cent of demand in Angola was met by locally produced textiles.[36] Encouraged by sales figures, the company also invested in cotton production in order to have guaranteed supplies for its textile factories. By 1973, moreover, the company had established another factory, Texmoque, in Nampula, Mozambique with an initial capital of 10,000 contos. This factory was to be supplied with second-hand machinery from bankrupt factories in Portugal and the company had arranged a ten-year line of credit from Portugal to purchase the machinery and transfer it to Mozambique. By 1975 the new factory's capital was up to 70,000 contos, most of it supplied by colonial enterprises.[37] These investors, who only two years previously had been confident of their ability to capture most of Mozambique's demand for textiles, were now nervously trying to ward off nationalization by the independent Mozambican state. For many in the Portuguese textile industry, all of these developments meant the end of their dominance of colonial markets. Exports of Portuguese textiles to

[36] See Sociedade Algodoeira de Fomento Colonial, *Relatório*, 1965–71; and Antonio Manuel Pinto, 'Comunicação', *Indústria do Norte*, 24: 642 (June 1973), 12.

[37] Universidade Eduardo Mondlane, Centro de Estudos Africanos, Instituto de Investigação Científica de Moçambique, *Capital privado, produtividade, e participação—um estudo da fábrica texmoque*', Relatório no. 80/2 (Maputo: UEM, CEA 1980), 6.

the colonies went from 15,088 tons in 1965 to 4,528 tons by 1973 (Table A5).

While markets and industrial structure were unsatisfactory for the majority of the industry during this period, equally as harmful to the industry were bureaucratic meddling, the rigid colonial cotton policy, and rising emigration. In all these areas, government policy contradicted industrial aims, and events conspired against attempts by the industry to pull itself out of the quagmire of the post-war years.

COTTON SUPPLY

The start of the colonial wars and the abolition of forced cotton production in 1961 initially affected the quantity of colonial cotton. Colonial cotton supply did not decrease precipitously, but after 1961 it was erratic. In 1964 the supply of foreign cotton outstripped the colonial supply for the first time since 1941. Following abolition of the cotton regime, concessionary companies and an increasing number of European growers, especially in Angola, began to mechanize large areas (see Ch. 12). From the mid-1960s with government support, the supply of colonial cotton rose again outstripping foreign cotton from 1967 to 1971. After that date, foreign imports again exceeded colonial cotton imports following legislation allowing the industry to purchase cotton freely.

As with quantity, neither the quality nor the price of colonial cotton proved attractive until the end of the 1960s. One importer pointed out in 1967 that cotton from the colonies was of such poor quality that it was unsuitable for use in the type of textiles now exported abroad.[38] Textiles produced for export abroad required the use of better quality cotton which the colonies could not supply in sufficient quantities. Government had eased restrictions on foreign imports of cotton, but now embroiled in the colonial wars, it maintained a commitment to enforcing the quota system until 1965 and to buying colonial cotton until 1972.[39] As a result, importers were often left with a contingent of colonial cotton while some industrialists bought better quality, foreign cotton from other

[38] Danilo Marques, *A problemática de actual situação do sector têxtil algodoeiro* (Lisbon: Sociedade Tipográfica, 1967), 18.

[39] *Diário do Governo*, Série I, 30 June 1965, Portaria 21,363.

importers; those industrialists who did not purchase better cotton risked having orders cancelled.

Along with the obligation to buy colonial cotton, moreover, the fixed price was maintained throughout the 1960s at a price higher than that of foreign cotton. The Stabilization Fund followed by the Export Development Fund and the Cotton Supply Fund after 1965 eliminated this difference for goods destined for export, but not for goods traded on the domestic market. Although colonial cotton was no longer allocated on a quota basis after 1965, the 1968 commission found that the CRCAR had actually forced cotton on importers at the higher price. Some of this lay on the dock unbought because modern mills preferred to import foreign cotton; many small mills could not afford it. The commission remarked, 'This raw material pricing policy was always imposed by the government and the industry never solicited nor advised the policy.'[40] In 1969 the Companhia Rio Ave pointed out that the difference between international cotton and colonial cotton was 3$00 escudos per kilo with the result that Portuguese textiles were more expensive and therefore difficult to sell in the colonies and abroad, once again exposing the internal market to dumping.[41] The Companhia da Fábrica de Fiação de Tomar declared:

> It is urgent that the supply of primary material, principally cotton, should follow the rules of international practice. With prices and quality inappropriate to production, the primary material compromises the entire manufacturing process, commercialization, and promotion of Portuguese textiles.[42]

By 1969 factories were demanding a liberalization of the cotton policy, and the system was dropped in 1972, ironically just when colonial quality and price were becoming more attractive due to increases in the world price and the mechanization of colonial cotton. Its role in managing the cotton supply apparently eliminated, the CRCAR was also abolished in 1972. In its place, the government created the Instituto dos Texteis (Textile Institute) an organization with more benign duties designed to serve not only the cotton textile industry, but also wool and synthetics, clothing and knitwear.[43]

[40] Comissão, *Relatório*, 22.
[41] Rio Ave, *Relatório*, 1969, no page.
[42] Tomar, *Relatório*, 1969, no page.
[43] *Diário do Governo*, Série I, 31 Oct. 1972, D-L 329/72.

SALARIES AND MIGRATION

If credit problems, cotton supply, and unstable markets placed some firms in financial difficulties and simply exasperated others, the interconnected development of inflation, emigration, and wage increases worsened the problems. As mentioned in the preceding chapter and in company reports, low wages had accounted for the low buying power in the country. This had been pointed out continually during the 1950s but to little effect since low wages also helped many Portuguese industries to be more competitive abroad. The government made few efforts to improve the standard of living so in the late 1950s when other European economies were expanding, large numbers of Portuguese people began to migrate in search of better wages.

Emigration affected the textile industry. Several companies referred to the loss of skilled labour and high labour turnover in their reports.[44] On a general level, the government tried to stem the flow by raising wages, but inflation made this ineffective. Whereas emigration led to high labour turnover and consequently, lower productivity, the wage increases and inflation led to higher production costs which in turn led to price increases and/or less money available for replacing machinery or paying interest on loans. One company, the Fabrica de Fiação e Tecidos de Ermezinde constantly referred after 1968 to high costs and lack of qualified personnel as well as the usual factors of unstable markets and erratic cotton supply to explain why its equipment was obsolete and had not been replaced.[45] Another, Rio Ave, asserted that low productivity and the poor quality of its textiles was due to its use of untrained apprentices recently hired to take the places of those who had emigrated, legally and illegally, to jobs in Northern Europe.[46]

The industry was now trapped by misfortune, by government inconsistency and favouritism, and by its own apathy or greed. Emigration, inflation, short-term credit at high rates of interest, unreliable markets, and unsuitable supply combined to strangle a deteriorating industry. Moreover, government policies had split the industry in two. Certainly there had been growth and an increase in exports, but these improvements benefited only a few.

[44] See Braga, Rio Ave, Tomar, Santo Tirso, and Ermezinde reports from 1964.
[45] Ermezinde, *Relatório*, 1968–71.
[46] Rio Ave, *Relatório*, 1969, no page.

Many small firms, plagued by high colonial cotton prices and cutthroat competition at home and in the colonies, struggled daily against the threat of closure.

CONCLUSION

Government plans to modernize and expand the industry only substantially benefited a small sector of the industry after 1958. Government aided the export sector through the Stabilization Fund, easing foreign cotton imports, and trade agreements, but did little to help those already operating under poor conditions in the internal and colonial markets. A divided industry was the result. Simultaneously, other government priorities, namely a commitment to the colonial cotton regime and to price stability contradicted its support for larger industries and worsened the predicament of many smaller, older firms. Such actions, unfortunately, had unintended consequences. The more modern sector of the industry retained an interest in colonial markets through its investment in local textile mills, but largely concentrated on exports abroad. Simultaneously, this sector resented the obligation to purchase colonial cotton. The export market demanded a steady supply of higher quality cotton that the colonies could not meet. By the late 1960s, the modern sector of the industry was calling for liberalization of policy and greater integration with Europe, and had a greatly reduced interest in the colonies. Once industry adopted the notion of liberalization, an interventionist government was a hindrance and the retention of colonies, an embarrassment. The other sector relied on the small colonial market where, by 1965, it had to compete with local mills partly financed by large metropolitan mills. At the same time, small mills could not afford high-priced colonial cotton. The predicament they found themselves in weakened their support for the government at the same time that it underscored their dependence on the colonies.

The industry had come full circle. More than forty years after a cotton policy had been implemented, an excess of contradictory, half-applied regulations had left the industry with its old problems of inflation, cutthroat competition, unstable and expensive cotton, and new challenges of emigration, colonial wars, and competition in foreign markets. This time, as before, the government had to balance other interests against the wishes of the textile industry. In the 1960s, the threat of loss, not simply of the Portuguese territories, but of its own political power, counselled maintenance of the colonial cotton regime. When the colonial cotton system was finally abandoned in 1972, it was a virtual admittance by the government that it was no longer able to control the liberalization at home and the liberation taking place in the African territories. Thus ended an era of government intervention in the textile industry; what began was a battle with its legacy.

12

The Collapse of Portuguese Colonialism

Whereas the period from 1946 to 1958 saw stronger ties between the metropole and the colonies established through increased trade, development projects, colonization, and administrative changes, 1958 to 1974 witnessed attempted economic integration and visible political deterioration between the two. Even as Portugal established greater trade relations with Europe, it did not abandon its colonies but tried to alter their place in the Portuguese economy. Import and export links continued, but the greatest area of interest was in establishing an economy with a single monetary unit and a single balance of payments—a Portuguese 'Common Market', eliminating tariffs and customs duties, and initiating investment and development projects in the overseas colonies to benefit the metropole. In much the same way as Britain and France had tried to do in the 1950s, Portugal appeared to be laying the economic foundations for a future neo-colonial relationship by constructing dams, devising hydroelectric and electrification schemes, and encouraging investment by Portuguese and foreign capital.

However, Portugal did not simultaneously prepare its colonies politically for independence by creating a 'class' of settler or African bureaucrats or businessmen to take over, and notionally favourable to its interests.[1] Efforts to do so in the mid-1960s came too late. By 1963 the Portuguese military were fighting wars in all three of Portugal's African colonies. As the wars dragged on, they shattered the political and economic configurations which the authoritarian regime had constructed in the metropole and in the colonies. On the morning of 25 April 1974, to the sound of Zeca Afonso's 'Grandola Vila Morena', a military coup effortlessly toppled the nearly 50-year-old dictatorship. Within a year, most Portuguese government officials, settlers, companies, and their capital, exper-

[1] Munro, *Africa and the International Economy*, 202–7.

tise, and possessions had evaporated from the former colonies—bequeathing to the new citizens of Mozambique and Angola only the bitter memories and rotten remains of colonial rule.

With the metropolitan textile industry embroiled in a crisis of over-production during this period, it was predictable that the price and supply of colonial cotton would be an area of intense conflict. Beset by other pressures and priorities, the government wavered between support for the domestic textile industry, incentives to cotton concessionary companies, and reform of the entire scheme. This balancing act appears finally to have benefited no one, least of all the Africans involved.

THE 1958 PRESIDENTIAL CAMPAIGN IN THE COLONIES

As the criticisms in the previous chapter on the colonies showed, by the late 1950s settlers had found a voice and begun to use it. According to Wheeler and Pelissier, 'economic freedom was more an issue than the vote',[2] but settlers did not refrain from using the ballot-box when available. During General Delgado's 1958 campaign for President, the two coincided. Despite government fraud, Delgado made a strong showing among settlers precisely because he advocated greater economic freedom for the colonies. While he did not support self-determination, he argued that the administration of the colonies was too centralized and that thirty years of Salazar's rule had resulted only in stagnation. Calculated to appeal to settlers, he focused on the slow development of mechanization, industrialization, and infrastructure in Angola and Mozambique.

The results of the elections in the overseas colonies must have badly shaken Salazar. Within and without the country, opposition forces had united to criticize sharply the *Estado Novo*. In Angola, with the President of the Industrial Association as campaign manager, Delgado won 33 per cent of the vote including the whole of the district of Benguela, the centre of commerce and industry there.[3] Although the results indicate settler dissatisfaction, it is

[2] Wheeler and Pelissier, *Angola*, 147.

[3] Bailey, 'The Political Process', in Abshire and Samuels (eds.), *Portuguese Africa*, 153; but see also Parsons, 'The Torment of Angola', 79. According to Parsons, Delgado won 20 per cent of the vote in Angola and 40 per cent abstained.

difficult to know how far settler resentment and organization would have gone or how willingly Salazar would have sought to appease them. Events in 1961 changed the entire context in which all interests operated.

THE COTTON REVOLTS

In spite of price increases and the encouragement of co-operatives, for many African cotton producers the attempts to improve the cotton regime during the 1950s were too little and too late. In late 1960 African cotton farmers working in Cotonang's Malange concession in the Baixa de Cassange killed their animals, left the fields, and refused to pay their taxes.[4]

The following January, Cassange farmers were supposed to be planting the cotton crop for the next year, but instead 'the farmers burned their seed, abandoned their tools, and tore up their identification booklets.'[5] The farmers then began attacking shops, administrative posts, and a Catholic mission but without bloodshed.

Three factors apparently combined to explain the occurrence of the strike. First, a 'messianic nationalist' movement led by António Mariano fuelled peasant hatred of the cotton regime and encouraged them to engage in the strike (which subsequently acquired the name, 'Maria's War' from Mariano's last name). Secondly, the area's proximity to the newly independent Congo may have led to contact with independence ideas and movements further encouraging peasants to strike. Thirdly, it has been claimed that the peasants were striking over falling cotton prices and poor working conditions.[6]

No doubt some contact with independence ideas had taken place, but there is no indication that an organized independence movement was behind the cotton uprisings. Moreover, strikes over prices and conditions were common in the cotton regime and the Portuguese had encountered them before; nevertheless, the Portuguese responded savagely to the protest. The Angolan govern-

[4] Henderson, *Angola*, 181.

[5] Ibid.

[6] Ibid. and John Marcum, *The Angolan Revolution* i. *The Anatomy of an Explosion (1950–1962)* (2 vols.; Cambridge, Mass.: MIT Press, 1969 and 1978), 124–5.

ment sent troops to Cassange and firebombed and napalmed the African settlement, reportedly killing hundreds or thousands of Africans, according to estimates.[7] At the same time, it brutally repressed an attack by Africans on a prison in Luanda.[8]

A month later, another strike by Africans occurred, this time in the coffee area. Protesting falling coffee prices, Africans ransacked and burned towns, administrative posts, and coffee plantations. Again the Portuguese responded by sending troops and bombing villages. In the six months of skirmishes which ensued between Africans and Portuguese troops, an estimated 20,000 Africans died; a total of 750 Europeans were said to have been killed also in the first three months of the revolt.[9]

Meanwhile in Mozambique, in June 1961, perhaps unnerved by recent strikes there and the trouble in Angola, government troops fired on a non-violent demonstration at Mueda killing about 600 Africans.[10] Accounts vary as to the purpose of the demonstration. On one account, peasants belonging to the Mozambique African Voluntary Cotton Society, a cotton co-operative, and other cotton farmers launched a demonstration in front of the residence of the district Governor to demand an increase in prices paid to cotton producers and a liberalization of agricultural policy.[11] Another account states that demonstrators were protesting against forced labour practices.[12] Yet another account claims that peasants, urged on by the Mozambique African National Union (MANU), a Mozambican independence movement based just across the border in Tanzania, launched a demonstration to demand liberation.[13] Some members of the cotton co-operative were members of MANU and indeed, a prominent member of the co-operative had been imprisoned on charges of sedition in 1959,[14] but there is no indication that a demonstration had been planned or violence anticipated. Mozambican nationalist movements had only recently formed since Tanzania's independence in 1961 and did not yet have an organized

[7] Ibid. i. 125.

[8] Wheeler and Pelissier, *Angola*, 175.

[9] Marcum, *The Angolan Revolution*, i. 140–7.

[10] Isaacman, 'Mozambique Cotton Cooperative', 17.

[11] Eduardo Mondlane, *The Struggle for Mozambique* (Harmondsworth: Penguin, 1969), 117.

[12] Mazrui and Tidy, *Nationalism and New States in Africa*, 133.

[13] Isaacman, 'Mozambique Cotton Cooperative', 17.

[14] Ibid.

following within Mozambique. Regardless of the demand, the government answered it with gunfire and the African cotton co-operative was banned that same year.

Though sporadic protests were brought largely under control and factionalism among the opposition prevented organized resistance until 1964 in Mozambique and 1966 in Angola, the Portuguese government tried to take political advantage of the upheavals and attacks in both countries to reassert its authority at home and in the empire. Capitalizing on renewed nationalist sentiment at home and among settlers, Salazar and the Overseas Minister, Adriano Moreira, presented Portugal as a martyred country wrongly attacked and condemned by all when it had only tried to perform its 'civilizing mission'.[15] Military intervention escalated in the colonies due to a succession of events in the period from 1961 to 1962. By 1962 Portugal was deeply embroiled in the colonial wars.[16] Under extreme pressure at home and in the colonies, 'the rebellion offered the government an opportunity to deflect and redirect latent popular discontent into a furious campaign to smash the rebelling African.'[17]

The policy of polarization rather than amelioration, the threat of loss and humiliation worked initially. Settlers ceased their criticism of metropolitan control, and opposition within Portugal divided over the future of the empire. Only one opposition group there dared to link liberation in the colonies with the defeat of 'fascism'. After preaching that Africa was a part of Portugal for nearly forty years, this was perhaps a major reason why Salazar decided to fight a costly and consuming war.

THE COLONIAL WARS: REFORM AND REPRESSION

The government's determination to keep the empire was backed by reform and repression in its two largest African colonies. First, the government granted greater political voice to the settlers. It created economic and social councils and Provincial Cabinets. In 1963 the Overseas Reform Act and the Organic Law of the Overseas

[15] See speeches by Salazar in the Bibliography from 1961 to 1963 and all of the speeches by Adriano Moreira.

[16] See Ch. 10 for the prelude to military intervention.

[17] Marcum, *The Angolan Revolution*, i. 154.

Provinces allowed for greater representation of Angolan and Mozambican settlers. Members of the Legislative Councils in each province were now to be elected rather than appointed and their number increased; but Councils did not sit for more than three months a year, giving them little effective power. The Governor of each colony and his committee were granted more administrative duties but the Overseas Minister still made all policy decisions.[18] Lisbon had relinquished little control but perhaps had found a way to blunt settler criticism and secure allies.

The government also made several changes in its political and social treatment of Africans. With respect to labour issues, in 1961 the forced recruitment of Africans for public and private work was dropped, and a Labour Inspectorate and a Labour Welfare and Social Action Institute created.[19] For cotton, Decree-Law 43,639 of 2 May 1961 abolished forced cotton growing and prohibited administrative authorities from any involvement in its development or commerce.[20] The JEAC, which had been responsible for intervention in and aid to the scheme, was abolished and reorganized into the Instituto de Algodão (Cotton Institute).[21] Similar changes took place for other crops such as cereals and coffee, but neither their formal duties nor their personnel appeared to change. The Cotton Institute continued to perform many of the same functions as in the past. Following the abolition of forced labour in the cotton scheme came the abolition of the forced rice-growing scheme in Mozambique.

For national and international consumption, Lisbon made a number of social reforms aimed at abolishing differential treatment of Europeans and Africans. These included the abolition of identity cards and the distinction between *assimilado* and non-*assimilado* status as outlined in Decree-Law 39,666, and the granting of Portuguese citizenship to Africans with the passage of Decree-Law 43,893 in 1961.[22] In 1964 the regime introduced compulsory primary education for everyone in the colonies. These revisions, both real and apparent, underscored the need for the *Estado Novo*

[18] Portugal, Agência Geral do Ultramar, *Organic Law of the Portuguese Overseas Provinces*, 1963 (Lisbon: Agência Geral do Ultramar, 1963).

[19] Antonio de Oliveira Salazar, *Declaration on Overseas Policy*, Broadcast 12 Aug. 1963 (repr. Lisbon: SNI, 1963), 21–2.

[20] *Diário do Governo*, Série I, 2 May 1961, D-L 43,639.

[21] *Diário do Governo*, Série I, 31 Dec. 1961, D-L 43,874 and 43,875.

[22] Henderson, *Angola*, 186.

to protect its own position by appearing to meet demands for reform.

Coupled with the political reforms were several economic reforms designed to appease international and settler criticism, while not compromising Portugal's overall economic objectives for the colonies. The Second Development Plan from 1958 devoted more public spending and investment to electrical power, ports, infrastructure, and communications, and expanded mechanization and colonization schemes.

With regard to trade and investment, the Export and Import Boards combined to form the Board of Trade (1956) in order to unite commercially the metropole with the overseas possessions; customs tariffs and duties were reduced gradually over a ten-year period beginning in 1961; the escudo was made the single unit of currency from 1963; and restrictions on transfers of funds to cover imports were lifted in order to allow colonial investors, importers, and exporters better access to lines of credit, and in order to facilitate financial transactions between the colonies and the metropole. Additionally, the regime also offered credit, loans, and technical supervision to private initiative in the colonies through the Economic Development Fund. The government further liberalized and encouraged private initiative by allowing foreign investment in Portugal and the colonies and vastly reducing Industrial Regulation in 1965. The Third Development Plan which followed these measures in 1968 gave 46 million contos to mineral exploration, transport and communications, and industry.[23] These goals were pursued alongside the policy of continued emigration to the colonies.

Such reforms were not an act of desperation but reflected the economic forces and political considerations operating in Portugal. With the financial-industrial oligopoly rapidly forging trade links with Europe, the government wished to expand investment opportunities in the colonies in order to prevent a shift of loyalties away from Africa. It was thought that through the creation of an escudo zone free from tariff barriers, economic activity between the colonies and Portugal would increase, the market would expand, and the standard of living would improve.[24] Equally, by establish-

[23] Brandenburg, 'Development, Finance', in Abshire and Samuels (eds.), *Portuguese Africa*, 227; and *III Plano*, vol. i.

[24] *Diário do Governo*, Série I, 8 Nov. 1961, D-L 44,016.

ing a free trade zone, Portugal wanted to accommodate European demands that it reduce protectionism and reinforce its claims that Portuguese territory also included its African colonies.

The results were not unsatisfactory but neither did they conform to the expectations of the metropolitan regime. Growth rates in Angola and Mozambique up to 1965 were around 4 per cent and were comparable to those in the rest of Africa.[25] In Angola, manufacturing comprised 16 per cent of gross national product by 1971 compared with 6.5 per cent in 1960.[26] The number of manufacturing enterprises more than doubled between 1961 and 1972 from 2,490 establishments to 5,561; many of these were financed by Portuguese companies and substituted for imports from Portugal and abroad.[27] No doubt partly as a result of the war, schools, hospitals, doctors, and streets in Angola and Mozambique all saw significant increases between 1958 and 1973 as Table 12.1 illustrates. Railroads in Angola were the exception. Emigration also proceeded apace. By 1970 the Portuguese population in the colonies equalled 440,000 of which 290,000 were in Angola.[28]

With regard to investment and even political support, many foreign companies also obliged the regime. The percentage of foreign capital in total annual investments in the colonies went from 0.8 per cent in 1959 to 10.5 per cent in 1962 to 26.7 per cent in 1966.[29] In exchange for huge profits and concessions, foreign investors 'concluded agreements with the government of Portugal to support her in securing peace and order' and donated money for constructing military barracks in the colonies.[30] Diamang was one of these. It had exclusive rights to over 1 million square kilometres to prospect and mine diamonds. Between 1961 and 1965, Diamang's profits rose 74 per cent. The Angolan government received a share of the profits amounting to 770 million escudos plus 72 million escudos for the company's share in 'the defence of the national patrimony'.[31]

[25] Brandenburg, 'Development, Finance', in Abshire and Samuels (eds.), *Portuguese Africa*, 227.

[26] M. R. Bhagavan, *Angola's Political Economy 1975–1985* (Research Report 75; Uppsala: Scandinavian Institute of African Studies, 1986), 16.

[27] Ibid.

[28] Newitt, *Portugal in Africa*, 164.

[29] Ferreira, *Portuguese Colonialism*, 28.

[30] UN, Office of Public Information, *A Principle in Torment*, ii. 52.

[31] Ibid. ii. 51–2.

TABLE 12.1. *Changes in Selected Indicators in Angola and Mozambique, 1958–1973*

	1958	1973
Secondary schools		
Angola	32	69
Mozambique	6	41
Hospitals		
Angola	77	197
Mozambique	78	99
Doctors		
Angola	217	561
Mozambique	171	289
Streets (km.)		
Angola	35,519	72,323
Mozambique	24,940	38,886
Railroads (km.)		
Angola	3,110	3,049
Mozambique	2,537	3,736

Source: Portugal, INE, *Anuário Estatístico (Ultramar)*, 1958–73.

In addition to substantial British and Belgian investments in raw material production already present in Mozambique and Angola, Britain, South Africa, West Germany, and the USA also invested in oil and mineral exploration in the two colonies during the 1960s. Whereas in 1958, 58,000 tons of oil were produced in Angola; by 1973, 7.4 million tons were produced mainly by foreign companies. Iron ore, which saw heavy investment by the Germans, increased from 106,000 tons in 1957 to 7 million tons in 1973.[32]

Growth, import substitution, and investment ironically did not have the desired effect of welding the metropole more closely together with the colonies, though it did satisfy other goals. As mentioned previously, colonial importers took advantage of relaxed exchange controls after 1963 not only to build up a deficit with Portugal but also with countries abroad. By 1971, with the economy subject to inflationary pressures and balance of payments difficulties, the regime then decided to reimpose strict exchange con-

[32] Bhagavan, *Angola's Political Economy*, 16.

trols on trade between Portugal and the colonies and to encourage the colonies to increase exports abroad. These measures indicate the resurgence of old obsessions: reducing the deficit, earning foreign exchange, and stabilizing the escudo. At the same time, they indicate new priorities, namely a desire to satisfy European demands by reducing protectionism and liberalizing trade.

Alongside the reforms came repression. From 1966 rebels began to work in eastern Angola and northern Mozambique and the war assumed a new intensity. By 1967 Portugal had 150,000 troops in Angola and 40,000 to 60,000 troops in Mozambique. On completion of duty, soldiers were offered free plots of land ranging from 10 to 34 hectares and loan subsidies during the first two years.[33] Under the scheme, about 4,800 soldiers had settled in Angola by 1966.[34]

Portugal's main method of counter-insurgency was to resettle Africans. This took two forms, one for defence, the other for development. One was a direct response to the insurgency and involved the resettlement of Africans who lived near borders or were susceptible to rebel influence. Its objective was defensive. These strategic hamlets or *aldeamentos* were rigorously controlled and were practically military camps. The other form, rural resettlements or *reordenamentos rurais* were more like the *concentrações* conceived in the 1940s: a method of improving the life of Africans by providing land, education, and social services.[35]

In the late 1960s, however, the Angolan and Mozambican governments had yet another reason for resettling massive numbers of Africans in the central and southern parts of Angola and Mozambique without making adequate provision for their maintenance—European colonization.[36] The settlement of Europeans served a number of objectives. First, additional settlers reinforced claims by metropolitan Portugal that the colonies were an integral part of Portugal. Secondly, the government hoped to gain support from fresh settlers by helping them to become prosperous. Thirdly, the locations of new settlements indicated that the government

[33] Patricia Wohlgemuth, 'The Portuguese Territories and the United Nations', *International Conciliation*, 545 (Nov. 1963), 5.

[34] Eugene K. Keefe, *Area Handbook for Portugal* (Washington, DC: USGPO, 1977), 100.

[35] Bender, *Angola under the Portuguese*, 160.

[36] Ibid. 181–94.

intended to use Europeans as buffer zones between Africans inside Angola and Mozambique and growing independence movements just inside their borders. In central Angola, land held by Europeans doubled between 1968 and 1970 and by the early 1970s, Europeans owned 60 per cent of the arable land in that country. Africans who had once engaged in prosperous cash-cropping in the area were forcibly moved.[37] In Mozambique, peasants were moved off the land in the southern and central parts of the country to make way for European colonization. Of the three objectives, the strategic one was badly served by resettlement. By 1969 over 1 million Africans had been moved and a Symposium on Counter-Insurgency in Lisbon concluded that this had probably increased the level of resistance among Africans in addition to its negative economic and social effects.[38]

COTTON REGIME UP TO 1974

It is interesting to go back and look more closely at the cotton regime during this period for it illustrated many of the changes taking place at the time. Initial revolts occurred in cotton areas in both colonies but surprisingly, cotton production continued until the end of colonial rule in 1974. Largely responsible for the stability of output were changes to the organization of production taken after 1958. In conformity with an emphasis on an incentive-based system adopted from 1946, the government continued to encourage the formation of African commodity producers during the 1960s and early 1970s. At the same time, it both complemented and contradicted this approach by fostering the growth of settler cotton production on small farms often taken over from Africans and using African wage labour. These strategies profoundly altered the cotton regime by 1974, but did little to ameliorate African discontent. Below we trace the changes in the system from 1958 and their social and economic impact.

It has been claimed that revolts in the cotton areas were related to falling cotton prices, a look at real producer prices suggests the claim is true (Table A10). African cotton producers had received a

[37] Ibid. 183; and Keefe, *Area Handbook*, 100.
[38] Bender, *Angola under the Portuguese*, 195–6.

price increase in 1955 but then world prices dropped below the colonial export price in 1958 introducing a moratorium on further price increases. As world prices remained below colonial export prices from 1958, nominal colonial export and producer prices remained stationary while real prices declined. By 1960 producers were protesting declines in their real income and seeking increases.

The government tried to counteract falling world prices and shore up colonial prices by implementing the Stabilization Fund. As noted previously, taxes and duties were dropped, and the government agreed to subsidize importers the difference between the prices of colonial and international cotton if the cotton was to be used in textiles destined for foreign export. It also agreed to subsidize exporters who sold goods made with colonial cotton to foreign markets. For those textile exporters, this at least pegged the colonial price to the world price even if it did not make the purchase of colonial cotton more advantageous for the entire textile industry.

The government appeared to have been seeking a number of objectives through this Fund: first, price stability, an overriding priority of the Salazar regime; secondly, prevention of the drain on foreign exchange even if it meant using state money for subsidies; thirdly, maintenance of the cotton regime. This third objective had a number of elements to it. Primarily, the government was, at this time, actively promoting its notion of unity between the colonies and the metropole. Cotton was a part of this integration policy, particularly in Mozambique. Additionally, cotton concessionary companies were well-organized and many were linked to powerful Portuguese financial and industrial interests. For example, the Banco Português do Atlantico financed Algodoeira Colonial Agrícola and Companhia Nacional Algodoeira; Espírito Santo supported the Companhia dos Algodões de Moçambique and the Sociedade Agrícola do Cassequel; and the Banco Nacional Ultramarino backed Cotonang.[39] These connections, combined with the companies' close ties to the JEAC (later the Cotton Institute) enabled cotton concessionary companies to exercise pressure on the metropolitan government to retain colonial cotton prices and illustrated their power *vis-à-vis* metropolitan textile companies. Finally, the government feared the social consequences of a drop in price

[39] See company profiles in Martins, *Sociedades e grupos*.

(which at any rate were not averted by a stationary nominal one). As early as 1955, the government had anticipated a possible negative reaction and made it clear that colonial prices would not be changed to match international ones. It stated in Decree-Law 40,405:

The fluctuations in international prices do not accord with the system of cotton production in our 'provinces'. Price stability is absolutely essential so that indigenous producers do not become discouraged by the drop in income in certain years—drops which appear incomprehensible due to their ignorance of international commerce. Long-term development measures can be executed without interruption or a loss of confidence as well. It is therefore natural that sometimes colonial prices will be inferior to those which can be obtained abroad, and other times they will be superior.[40]

When the international price fell below that of the Portuguese price in 1958, the government therefore chose not to lower the price of colonial cotton, despite the attractions for the textile industry of expansion into export markets and the possibilities of integration with Europe. Stability and the encouragement of colonial production took precedence over the benefits of lower prices to the industry. Great efforts had been made in the post-war period to increase and improve the cultivation and ginning of colonial cotton, which now averaged over 80 per cent of total cotton imports; lowering prices to suit the textile industry would have been counter-productive. The majority of colonial cotton was still harvested by hand extremely inefficiently, and still required much administrative surveillance; a drop in price would surely have led to a drop in production and therefore a drop in profits for cotton concessionary companies. These considerations took precedence over the hardships which might be inflicted upon metropolitan textile factories as a result.

The need for a fund to subsidize the price of colonial cotton nevertheless exposed deficiencies in the cotton regime. Primarily, price was the major factor which tied the industry to colonial cotton. Once the price was above international prices the industry lost interest. Particularly with regard to that sector of the industry which was modernizing and competing in foreign markets, colonial cotton also did not offer quality advantages once the international

[40] *Diário do Governo*, Série I, 24 Nov. 1955, D-L 40,405.

price of foreign cotton had dropped. Competing abroad required better quality cotton at competitive prices. As in the 1930s, colonial cotton was once again unable consistently to meet the needs of the domestic industry in terms of quantity, quality, and price.

Secondly, the fact that government subsidies were now required to make Portuguese colonial cotton attractive made a mockery of the myth of 'cheap' African labour. In a world which was achieving high productivity through mechanized cotton cultivation, picking by hand was expensive. As a result, several companies and the JEAC started to experiment with mechanized cotton cultivation at this time.[41]

These realizations coupled with the cotton revolts and the commitment to development, private initiative, liberalization, and emigration prompted the government to consider different approaches to forced labour and the concessionary system, even while it applied the Stabilization Fund. Not only did the government revive its interest in altering the regime from one of force to one of incentives—a path it had only timidly taken in the 1950s, but also it began to consider settler farms as an alternative to African peasant production.

In spite of its violent reactions to cotton protests in Angola and Mozambique and its determination to maintain prices at the old level, the government abolished forced labour in 1961 and granted price increases to both concessionary companies and producers that same year. The price increases were intended to act as incentives to counteract possible labour shortages arising from the abolition of forced labour. Cotton concessions worked on a zone system whereby they ginned cotton that had been produced in their zone of jurisdiction. With labour now theoretically 'free' and administrative authorities prevented from interfering in the regime, potentially, companies were tied into zones in which no cotton might be grown at all. The rise in producer prices was intended to give an incentive to Africans to remain in the cotton zones.

In many areas, the price rise proved an inadequate incentive to Africans to maintain production. Sena Sugar Estates' cotton production dropped dramatically from 375 tons in 1960 to 14 tons in 1962.[42] Total production in both Mozambique and Angola dropped

[41] *BGU* 32: 370 (Apr. 1956), 106; and see e.g. statements by Cotonang, 1961–2, 1965–74; Algodoeira Colonial Agrícola, 1966, 1968; Companhia Colonial do Buzi, 1959, 1961.

[42] Vail and White, *Capitalism and Colonialism*, 384.

initially after the 1961 legislation and several companies reported poor quality crops and low profits. To underscore their fears about the future of cotton, several companies collectively decided to cease operations using the facility conceded by Article 8 of Decree-Law 40,405 of 1955 allowing for cessation of cotton activities with one year's notice. Unless the government altered the rest of the regime, the companies would cease cotton operations a year from 19 December 1962.

Decree-Law 45,179 of 1963 abolished the cotton concessionary zones. Henceforth, trade in unginned cotton was to be exercised by private initiative under competitive conditions. Minimum producer prices were still fixed (but not maximum) and buying and selling still conducted in government-run markets but companies now bargained and competed for African or European-produced cotton. Companies could buy unginned cotton and then gin and export it themselves, or agree to gin cotton for a fee. The advantage of this system for producers was that they could keep their own cotton, pay a fee for ginning, and sell it to exporters. For companies, the advantage was that they could bargain for as much cotton as their ginning capacity would allow.

Although many observers have seen the abolition of the cotton regime as an attempt by Portugal to reform labour practices and to bow to international pressure, specifically it appears to have been a reaction to the shortage of hands once forced labour was ended. Generally, it formed part of the state's new programme for the colonies in the 1960s. The trouble with the concessionary system was that companies could not compete for cotton production and were unable to expand because they were restricted to designated areas of operation. Within these zones, companies had to make do with the available land and labour even if neither were suitable or sufficient. What made the scheme possible was government support for the recruitment of cheap labour and government participation in the scheme. Once that was dropped, the scheme was unprofitable.

The legislation abolishing cotton concessions changed the organization of production and trade and gave the companies the ability to operate in free labour, trade, and property markets—probably the nearest to capitalism that Portuguese colonialism ever got. In a free labour and property market, companies could mechanize directly or compete for cotton produced by Africans. Equally important, the government could pursue European settlement in areas previously controlled by companies. In a free market, companies could compete for cotton production according to ginning capacity. So while abolition of the cotton concessionary regime may have been due in part to international pressure, it appears that it was an admittance also that a concessionary system could not work without forced labour and the cotton regime could not be productive with it. Those companies which had threatened to cease operations were allowed to opt out immediately; those who had not were given until 1966 to prepare for the change.

There are two caveats which probably helped to soften the transition from concession to industrial/commercial enterprise. First, producer and export prices for colonial cotton were raised in 1963. In 1963 and 1964 the colonial export price was 6 escudos higher than the world price, providing a strong incentive to companies to stay in the cotton business. After 1964, although the export price dropped from 22 escudos to 19 escudos, it was still between 3 and 4 escudos higher than international prices until 1969. Thereafter, due to the adoption of international standards of classification and the abolition of fixed prices in 1972, Portuguese prices began to conform to international ones but these were increasing after 1969. American middling, for example, went from 15$76 in 1969 to 35$92 in 1973 (see Tables A7 and A8). Secondly, Decree 45,179 of 1963 also instructed the new Cotton Institute to help supervise and develop cultivation thereby partially contradicting the earlier law preventing administrative involvement. Several cotton companies refer to and praise the Institute for aiding the cotton scheme in the 1960s.[43] Though the Cotton Institue was probably never as involved as the JEAC with compelling labour for work in the cotton fields, it did play an active role conducting research, classifying cotton, setting prices for seed cotton, organizing markets for the purchase of seed cotton, ginning cotton that companies did not have the capacity for or in areas where companies did not operate, and particularly, aiding new settler cotton farms through the provision of land, infrastructure, tools, advice, and credit.[44]

In light of the changes to the scheme brought by the new law in 1963, it is interesting to see the response by companies. In

[43] See statements by Companhia Colonial do Buzi, 1964–5; Sociedade Agrícola Algodoeira, 1966; Algodoeira Colonial Agrícola, 1964 and 1970; Cotonang, 1964.

[44] Universidade Eduardo Mondlane, Centro de Estudos Africanos, 'Cotton Production in Mozambique: A Survey 1936–1979', *Relatório*, no. 81/1 (1981), 18 and 21 (hereinafter UEM, CEA, 'Cotton Production').

Mozambique the result negatively affected most companies because they made little effort to mechanize production, upgrade their gins, or increase ginning capacity. Equally, cotton producers took advantage of legislative changes to abandon the companies and gin their cotton instead with the Institute, or to market it independently. Most cotton companies in Mozambique thus struggled throughout the 1960s. The Companhia da Zambezia dropped its concession in 1964; the Companhia Nacional Algodoeira repeatedly lamented reductions in the amount of cotton bought from producers and its poor quality throughout the 1960s but it did initiate mechanization and began to buy cotton from settlers in the early 1970s. The Companhia Colonial do Buzi, while it initially seemed pleased with its new industrial and commercial duties, later referred to cotton laying on the dock in Leixões (near Oporto) unbought by factories because of its poor quality and high price. By 1970 the company reported that production had dropped. SAGAL also reported very poor quality and decreases in quantity after appearing pleased by the new system, yet it began to engage in direct mechanization and spoke of remodelling its factory in the early 1970s.[45] When Mozambique became independent in 1974, SAGAL's report for that year promised to make every effort to contribute to the country's development.[46]

In spite of the difficulties experienced by the cotton companies, total production from Mozambique did not decline during the 1960s. After a large drop in 1961, Mozambique's total production started to rise so that by 1967, Mozambique was actually exporting 119 tons of cotton abroad in addition to 37,907 tons to Portugal. By 1973 Mozambique exported 60,861 tons of cotton, 27,151 tons to markets abroad.[47]

Two developments explain why Mozambique's production increased even after forced labour and the concessionary system were abolished: the existence of African commodity producers and the growth of European cotton farms. With regard to the first, we noted in Chapter 9 that peasant production was beginning to stabilize in the 1950s. Hectarage and the number of peasants did not see the huge fluctuations of the 1940s. After 1961, although

[45] See company reports by SAGAL, 1965–74; Companhia Nacional Algodoeira, 1962–74; Companhia da Zambezia, 1964; and Buzi, 1964–70.

[46] SAGAL, *Relatório*, 1974, no page.

[47] OECD, *Trade by Commodities*, Series C, 1967–73.

total peasant production dropped from 545,000 peasants in 1967 to 284,000 by 1973, nevertheless 'Groups of peasants using advanced techniques and specializing in cotton had begun to emerge in certain areas in the 1960s.'[48] These peasants saw the share of the export price accounted for by producer prices go from 17 per cent to approximately 37 per cent as producer prices increased from 3$80 in 1964 to 8$00 by 1973.[49] Also, they did not have to market their cotton via the old concessionary companies; instead, many formed co-operatives or chose the Cotton Institute and independent traders through which to gin and market their cotton.

Along with the stabilization of producers, an increase in output, and the creation of commercial relations came differentiation among Africans. In case studies of three different regions in Mozambique by N. S. Bravo, incomes from cotton in 1961 ranged from 100 escudos to 6,777 escudos. Out of a total of 15,588 producers surveyed, 477 or 3 per cent made over 1,800 escudos return on their cotton crop, an average of 12 escudos per day (taking 150 days devoted to cotton production) more than the average daily salary of 7 escudos for plantation work and evidence that, given average outputs per hectare at least 3 per cent of cotton farmers had become fully commercialized and had become the *agricultores* outlined in the legislation of 1948. Using tractors and insecticides to increase output, the transformation of peasants into commodity producers and their differentiation intensified during the 1960s and early 1970s.[50]

Yet checking this advancement of African commodity producers was the arrival of European settlers:

> Settlers were able to appropriate prime cotton land, or to pay only nominal prices for it; often the land had been cleared by the peasantry for *picadas* or for the *concentrações algodoeiras*; and of course, the settlers clustered in areas where good roads provided the best access to the ginning factories.[51]

Settlers did particularly well in Nampula where, conveniently, they acted as a buffer zone against the advances of FRELIMO, the

[48] UEM, CEA, 'Cotton Production', 15.

[49] Export price for 1973 is based on an average from Mozambique, since colonial cotton was subject to international pricing and classification after 1971. For producer and export prices, see UEM, CEA, 'Cotton Production', 19; and *Folha Textil*, no. 5 (Dec. 1976), 26.

[50] Bravo, *A cultura algodoeira*, 198–200.

[51] UEM, CEA, 'Cotton Production', 15.

resistance group acting with increasing effectiveness in northern Mozambique from the mid-1960s. European cotton farmers employed seasonal African wage labour and enjoyed considerable aid from the Cotton Institute to increase output by the early 1970s. Whereas in 1968/9 only 27 per cent of production in Mozambique was accounted for by settler farms, by 1973/4, settlers produced 67 per cent of total cotton output.[52] Again, while some settlers ginned and marketed their cotton with the companies; others chose the Institute or handled it themselves.

In Angola too, cotton output increased during the 1960s. High prices for colonial cotton, increasing demand from local textile mills, and the construction of another textile factory in Angola probably offered sufficient incentives for those involved in its production. Like Mozambique, Angola also witnessed the transition of African peasants into commodity producers and the growth of settler farms. Although the total number of African producers in Malange (the area of the greatest cotton production in Angola) declined from 31,379 in 1959 to 16,908 producers in 1970,[53] those Africans who stayed in cotton production saw substantial increases in output due to state support for mechanization, insecticides, and soil studies.[54] In 1970 the average return per producer equalled 2,931$00 escudos, an amount which far exceeded the return from other crops in Angola.[55]

At the same time, many settlers, backed by government finance, started to engage in cotton production in this period and increases in Angola's total cotton output in the late 1960s were said to have been due 'mainly to the higher production from the commercial growers' (i.e. Europeans).[56] By 1971, 73 per cent of total cotton production in Angola was by Europeans.[57] Moreover, companies were not undermined as in Mozambique. In Angola, several companies mechanized and improved their gins thereby increasing output.

Colonial company reports from Angola confirm an increase in

[52] Ibid. 55.

[53] Fernando Pacheco, 'Agricultura e sociedade rural na Angola dos anos 60: O caso de Malange', II Colóquio Internacional em Ciências Sociais sobre a Africa de Língua Oficial Portuguesa, Guiné-Bissau, 19–26 Nov. 1991, 47.

[54] Ibid. 31.

[55] Ibid. table VI.

[56] Standard Bank Group, *Annual Economic Review: Angola*, 1968, 3.

[57] Ibid. 1972, 7.

cotton output, European cotton production, and mechanization during the 1960s. After an initial period of confusion, Algodoeira Colonial Agrícola of Angola reported increases in the amount of cotton it was ginning and the construction of a new ginning factory at Novo Redondo in 1966.[58] It spoke of expansion and high outputs throughout the late 1960s and referred to the increase in European production also at this time. In 1968 this company became Algodoeira Agrícola Angola backed by Banco Português do Atlantico and Banco Comercial de Angola. It had ginning facilities with a total capacity of 22,000 tons unseeded cotton and was one of the four largest firms in Angola.

Cotonang after 1963 also spoke of expansion. Now that supervision of African production was handled by the Institute, Cotonang limited its activities to buying and ginning African-produced cotton in over sixty-seven markets in the Lower Cassange area as well as from European producers who had recently located in Angola. Additionally, Cotonang instituted direct mechanization on its own land, but this was limited. In 1965 the company declared:

> We were greatly satisfied to see that the example set by Cotonang with regard to the direct and mechanized cultivation of cotton, has been followed by numerous European farmers who have settled in the Lower Cassange. They have tried to develop their cotton farming operations, and in this way have contributed, as we have been doing for many years, to the European population and agricultural economic development of that important region of the Malange district.[59]

By 1969 mechanized production by 230 European farmers on 16,120 hectares produced 21,705 tons of unginned cotton compared to traditional production by 24,520 Africans on 14,200 hectares which produced 5,192 tons for Cotonang. Overall in Angola, production of cotton rose during the 1960s as mechanization increased; in 1972, 70 per cent of the cotton Cotonang bought from Africans had been mechanically harvested.[60] Nor was this cotton of poor quality either. In 1971 the Portuguese government began to classify colonial cotton according to international standards of length and strength; by 1972 the government took advantage of colonial improvements in cotton quality and output and a rise in interna-

[58] Algodoeira Colonial Agrícola, *Relatório*, 1966.
[59] Cotonang, *Relatório*, 1965, 6.
[60] Ibid. 1969, 1970.

tional prices to release the textile industry from its obligation to purchase colonial cotton (ironically, just when the quality had improved).[61] By dropping the metropolitan obligation to purchase colonial cotton, the government also withdrew its subsidization of colonial cotton, and fulfilled some of its obligations to the European Community by eliminating protectionist trade in cotton. At the same time, the government took the opportunity to earn desperately needed foreign exchange by trading colonial cotton abroad.

In some cotton areas, it seemed that Portugal's goals of incentives, European colonization, and mechanization were being realized. Even if the textile industry could not appreciate or even benefit from this development by the late 1960s, some Africans, settlers, and cotton companies did. As late as 1972, with the wars raging nearby, the profits of Cotonang equalled 12,475 contos. Even in 1975, when independence had been granted, this company quietly switched horses with the hope of retaining what was now an organized, mechanized, and profitable concern. In its report for 1975, in spite of its facilities having been looted and ransacked after independence had been declared, Cotonang did what any business with heavy investment and favourable profits would do, it supported the victors:

> Predictable in 1974 as a consequence of the historic turn which occurred in Portugal, the independence of Angola, proclaimed on the 11 of November 1975, completed a long journey of struggle by its citizens and gave to the Angolan people the longed for opportunity to construct their future as a capable and free people.[62]

Had independence been handled differently by the Portuguese and the liberation movements, there is little doubt that companies such as Cotonang, for better or worse, would have remained in Angola and Mozambique.

Changes in the cotton regime during the 1960s illustrate a number of points about the connections between Portugal and its colonies and the operations of Portuguese capital in those colonies up to independence. First of all, the Portuguese government did make some attempts at economic reform, both before and after 1961 and seemed to be moving rather rapidly towards capitalism throughout the 1960s. Capitalism brought risks and benefits as the

[61] *Diário do Governo*, 13 May 1971, D-L 202/71 and 6 June 1972, D-L 189/72.
[62] Cotonang, *Relatório*, 1975, no page.

failure by several Mozambican cotton companies to adapt to the challenges shows. The disruptive effects of European settlement on African producers also indicate that at various junctures, the government undermined its own strategies. Secondly, the rapidity with which mechanization occurred in many areas reveals that capital was available under certain circumstances and not lacking as many have claimed. Thirdly, apart from the injustice and social disruption caused by forced labour and the concessionary scheme, it appears that it had also become economically disadvantageous, suggesting that the decision to abandon it was as much on the grounds of its inefficiency as of its social effects, or its vulnerability to international criticism. Increases in output in both Angola and Mozambique after switching to a scheme based on African commodity production and European settler farms give support to the view that the scheme adopted in the 1930s brought few rewards by the 1960s. Fourthly, once some companies began to profit from mechanization, they wanted and intended to stay—even after 1972 when the quota system was dropped and the metropolitan domestic textile industry was allowed to buy cotton when and where it liked. Fifthly, moves by Portugal to integrate with Europe and stay in Africa were not economically mutually exclusive: the introduction of settler cotton farms, the encouragement to African cotton producers, and the activities of several cotton-ginning companies in the late 1960s suggest that Portuguese economic interests were indeed trying to do both. What was impossible was for Portugal to join the EEC and improve its international reputation while fighting a colonial war, hence the increasing support for a neo-colonial solution by many Portuguese businesses by the late 1960s.

Economic reforms, increased foreign investment, improvements in infrastructure and electricity, increased trade with Europe, and economic integration with Africa did not go far or fast enough, however, and were not accompanied by substantial political reforms. Independence was out of the question while the regime maintained farcical elections, censorship, and political repression at home. Settler demands were ignored in the face of continued centralized control from the metropole, and African protest, answered with gunfire. Moreover, this political and economic situation was mirrored in the metropole as the government attempted to cater to the demands of the financial-industrial conglomerates while depoliticizing the majority of the Portuguese. After the departure of Salazar, it was this contradiction between purported flexibility in the economic arena and rigidity in the political field which brought the collapse of the regime. Confronted with growing trade and investment in Europe and violent struggles for independence in three colonies, the regime's mixture of economic liberalization and political authoritarianism finally proved incompatible.

Conclusion

During nearly half a century of association, the triad of regime, textile, and colonial cotton interests rarely coincided and frequently clashed. Policies were sometimes arbitrary and their effects, unintended. The regime often valued its own objectives above the interests of colonial cotton companies and African producers or the needs of the Portuguese textile industry. The colonial cotton system placed hardships upon producers and exasperated manufacturers. State intervention was pervasive; the views of those affected by government policies were neither fairly nor adequately represented.

Contrary to what has been commonly assumed, the Salazar and Caetano Governments did not always act in favour of the textile industry nor did colonial cotton consistently benefit it. Instead, contradictions and uneasy compromises amongst the regime, textile manufacturers, and colonial cotton companies punctuated the period from 1926 to 1974. The maintenance of protected markets in the colonies for metropolitan textiles, the existence of colonial cotton prices which were lower than international prices from 1941 to 1958, subsidies after 1959 for textiles destined for foreign markets, and the preservation of low wages depict only one facet of the regime's relationship with the Portuguese textile industry. The regime also passed modernization legislation in 1946 which was biased against small factories, and it maintained colonial prices which were higher than those quoted on the international market after 1958. It established total and individual quotas of colonial cotton to be purchased by the industry, and ignored textiles in favour of modern conglomerates after World War II. Government-run organizations formulated, implemented, and enforced policy decisions without consulting the majority of manufacturers. These actions also describe the relationship between the regime and the domestic textile industry.

Regime policies and objectives similarly influenced the relationship between the textile industry and the colonies. In spite of lukewarm enthusiasm by textile industrialists for colonial cotton in the 1930s; by the 1940s, they welcomed the availability of cheap colonial cotton at a time when supplies were erratic and international prices, high. Simultaneously, wartime plus highly protectionist policies allowed manufacturers to capture colonial markets and even to expand abroad. Inexpensive cotton and inflated export values brought high profits to mills during World War II. This was the period of the most advantageous links between the colonies and the industry.

In the post-war period, the synchrony between the industry and the colonies began to disappear because the regime was unwilling or incapable of reconciling a number of conflicting objectives. On the one hand, it was determined to modernize a sector of the industry. On the other, the quality of colonial cotton was unsuitable for the new machinery and new markets desired by larger, more modern factories that domestic policy had so favoured. Once these mills moved into European markets after 1958, the price and quality of colonial cotton became an area of contention between the regime and more modern factories. Concomitantly, having invested in local colonial textile mills, large manufacturers lost interest in supplying colonial markets.

For smaller factories the connection with the colonies served a different purpose. The post-war drive to modernize placed emphasis on specific, ideal dimensions which did not apply to the majority of textile factories in Portugal. Hobbled by biased legislation, poorer, smaller factories used low wages, colonial markets, and most of all, the imperfect operation of the colonial cotton quota system to undermine the regime's modernization policy. Indeed, the government legally abandoned the policy in 1955 once its social consequences became clearer. Struggling to survive, these smaller, dilapidated firms were then confronted by colonial cotton prices which were higher than international prices after 1958, lack of capital for investment in new equipment, emigration, and a contracting colonial market due to the growth of local textile mills financed by large metropolitan textile companies. In the 1960s, the regime was embroiled in a colonial war and stubbornly refused to alter the cotton system. Poor quality, expensive colonial cotton continued to be imported into Portugal throughout the 1960s now to the dismay of smaller, deteriorating factories too. Only in 1972 did the regime drop the system in accordance with new priorities, ironically just when colonial cotton had improved.

If the colonial cotton system presented difficulties for the metropolitan textile industry, equally it caused confusion in the colonies. The implementation and maintenance of the system provoked African resistance and migration, and squabbles amongst companies vying for African labour. Repeatedly, colonial administrators, chiefs, and overseers intervened to coerce labour into the scheme or mediate between companies. The constant vigilance of the JEAC was required to ensure that the seed was planted, harvested, and sold, and that no Africans died from hunger in the process.

After the war, many cotton producers responded to incentives—from price rises to settled African households—but force still remained a major feature of the system. Subsidies for colonial cotton after 1958 and revolts in the cotton areas in 1961 verified that 'cheap' African labour was both unprofitable and angry. Anxious to retain the colonies, the regime adjusted the scheme again, if only to maintain its own survival. It abolished forced labour and the concessionary system and opted instead to bring the logic of the market to Angola and Mozambique. As with policies in the metropole, the approach contained serious defects. The encouragement of African commodity producers conflicted with the drive to settle Europeans. Both needed land, tools, labour, and advice. Through revolt and ultimately, triumph, most Africans demonstrated their rejection of these and other flawed reforms of the colonial system.

Examination of the operation of policies and the interests affected by them in the period from 1926 to 1974 provides several insights into the development of the Portuguese textile industry, the system of colonial cotton production, and the nature of the regime. Government regulation combined with colonial cotton supplies and markets to shape profoundly the development of the textile industry. The bureaucratic contortions necessary to maintain stable colonial cotton prices, a guaranteed purchase of colonial cotton by the metropole, and protected markets reveal that autarkic policies are difficult to pursue in an industry which is vulnerable to international price trends and dependent on foreign exports. In two instances, the maintenance of stable colonial prices and the quota system nearly undermined sectors of the Portuguese textile industry: the first was when small industry used cheap colonial cotton and the quota system to defy a modernization policy designed to hasten their demise; the second was when higher colonial cotton prices and quotas were maintained in the face of falling international prices and the need for higher quality cotton. Compared with strangulating regulations and contradictory policies, *laissez-faire* and free trade policies would seem healthy alternatives.

No doubt Portugal's colonial cotton policy has left damaging and possibly permanent scars on the economies and peoples of Mozambique and Angola, but from nearly half a century of mismanagement and exploitation the empire exacted its price on the metropole as well. The profound impact that colonial cotton had upon the textile industry in the 1950s and 1960s suggests that while colonies may be exploited to the benefit of the metropole, empire can also undermine metropolitan government and industry. The availability of colonial cotton during the 1940s allowed the industry to capture markets during World War II; but also its erratic supply, poor quality, and high price were contributing factors to the prolonged crisis of the industry in the 1950s and to its division by 1974. Trapped in the colonial cotton system by a state anxious to cater to colonial interests and protect its own treasury, most factories had little control over the price and quality of their manufactures and their own growth and development. After 1958, part of the industry was no longer interested in colonial cotton; the other part could not afford it. For exports, modern mills were integrating with Europe; the rest, mostly small, old, and dilapidated, were left dependent on a stagnant domestic market and a contracting colonial one.

The fact that the administration of the colonial cotton policy nearly undermined the textile industry allows us to draw some preliminary conclusions about the nature of Portuguese colonialism. More research is needed on the functioning of the corporative system in the colonies, the different economic roles that Angola and Mozambique played in relation to the metropole, and the role of the colonies in the ideology as well as the economic goals of the *Estado Novo* and its successor, the *Estado Social*; but investigation of the regime's colonial cotton policies suggests that like other imperial powers, it had similar motives for establishing a regular supply of colonial raw materials for metropolitan use—to save on foreign exchange. It implemented a cotton regime similar to those in force in other colonies, encountered many of the same challenges, and gave Portuguese textiles preference in the colonial market just as Britain provided colonial markets for its textile industry.

While economic motives were similar, the policies show institutional and political differences from the policies of other colonial powers. The Salazar Government transferred corporative institutions to the colonies, and with them, their drawbacks: excessive bureaucratization, corruption, favouritism, and arbitrary use of power. The conflicts between the JEAC and colonial administrators, other companies, African producers, and even metropolitan textile industrialists, illustrate the disadvantages of a system of controls and excessive force. The adoption of incentives and the shift to market principles—a move towards capitalism—reveal that the regime, too, realized the inefficiency of force and was influenced by the responses of Africans. However, capitalism (along with new priorities generated by the war and Portugal's turn towards Europe) brought its own contradictions, thwarting the regime's objectives.

Moreover, even when the cotton policy had exhausted its purpose for metropolitan industry, the regime retained it. Mercantilist and political aims of the *Estado Novo*, the declining influence of the textile industry in the Portuguese economy, and the growth of powerful colonial cotton interests prevented the abandonment of the colonial cotton system even when its price was higher than the world price, its quality was unsuitable, and its treatment of African producers was criticized. These conflicts and problems explain why Portuguese colonialism is often perceived as brutal, chaotic, and unsuccessful.

The pursuit of government objectives in spite of their deleterious effects on industrial or cotton interests, the biased treatment of sectors within the textile industry, the favouritism accorded to modern financial-industrial conglomerates as opposed to more traditional industries, and the use of unrepresentative, government-run organizations such as the CRCAR or the JEAC to carry out the regime's goals leave little doubt as to the nature of the regime. Interventionist and authoritarian, it was also arbitrary, illogical, and probably corrupt in many of its actions. In the relationship between the regime and this industry at least, representation through the corporative system was a myth. Aside from larger, more modern textile firms who had informal and indirect contact with the government, most factories had little influence with government officials. Decisions were not tailored to the specific needs of

450 mills operating under extremely diverse conditions, but according to bureaucratic preferences, political self-interest, or technological criteria.

These findings substantiate the conclusions reached by other, recent studies (referred to in the Introduction) on the authoritarianism of the *Estado Novo* and suggest that ideological pronouncements and restricted readings of legal texts do not provide an adequate guide to the representation or the suppression of group interests in a non-democratic system. Moreover, the contradictions in domestic and colonial policy, the regime's arbitrary behaviour towards textiles, and its increasing intervention provide evidence of a growing disjuncture amongst the political, economic, and ideological bases of the authoritarian regime—the evolution of which has been suggested by the periodization.

From 1926 to 1936, the *Estado Novo* was able to restore stability to Portuguese society by arbitrating amongst economic élites, appeasing labour through a carrot and stick approach, and reasserting Portuguese imperial strategy. Moreover, it offered an ideology which was not only fashionable elsewhere in Europe but also included beliefs and principles which appealed to most Portuguese: the glories of an imperial past, the tenets of Social Catholicism, the notion of social harmony. In addition, the conflicts which had taken place during the Republican period as well as the effects of the Depression demanded extensive state intervention which the weakness of economic élites allowed. The regime's encounter with the textile industry exemplified the general changes in a specific context. Markets were secured, and Industrial Regulation implemented. In the colonies, moreover, metropolitan control was reasserted and an attempt made to get the colonies to serve the metropole. At this time, then, ideological, political, and economic motives of the *Estado Novo* coincided and the interests of economic groups and the regime overlapped.

From 1936 to 1946, the *Estado Novo* used its newly acquired economic and political leverage and the advantages afforded by the Spanish Civil War and World War II to intervene further in Portuguese society, to shift its support to industry, and to tie the colonies further to the metropole. Under cover of the war, the state actively intervened to reduce competition and pacified metropolitan industries with protectionism. It used legislation, bureaucratic organizations, and coercion to obtain cheap and necessary raw materials for metropolitan industry. By 1946 the regime exercised considerable control over all aspects of Portuguese life.

Having benefited from state patronage as well as from economic opportunities brought by the two wars, powerful financial-industrial conglomerates emerged in the post-war period. Reacting to the changed international environment as well as to the economic growth in its own country, the regime adopted a technocratic, developmentalist language to divert international attention away from its highly interventionist, unrepresentative nature and to address the new economic configurations in the country. These changes indicate a state that is responsive to pressure and capable of changing strategies, yet without sacrificing its own priorities. It passed development plans but within the strict limits of a balanced budget policy. It tried to foster growth without upsetting the balance of payments. With regard to industry, the regime shifted between support for modernization and heavy-handed interventionism. In colonial affairs, the regime advocated development and industrialization but still expected the empire to serve the metropole economically and ideologically.

After 1958 this position became untenable; interests diverged as the regime evolved. The colonial wars drove a wedge between the regime's neglect of traditional industries and its liberal overtures to big business, between its links with the colonies and growing integration with Europe, between the conservatives and technocrats in the government, and between ideology and reality. Emigration, inflation, and widespread opposition widened the gap. In addition, once the colonial wars began, the economic and political actions of the regime contradicted each other. Economically, the *Estado Novo* had laid the foundations for a future neo-colonial role, but these had not been accompanied by substantive political changes such as the promotion of settler aspirations or the creation of a class of Africans whose priorities might be notionally coincident with its own. Politically, the regime continued to equate the concept of the Portuguese nation with the existence of colonies. Having built its authority upon a mythical image of the Portuguese empire, the regime found itself fighting for its own political survival once this tenet of its authority had been challenged. The *Estado Novo* did not fight to retain its empire because of economic weakness, but because of the fragility of its own political legitimacy—hence the longevity of Portuguese colonialism and the violence required to crush it.

Clearly a single case-study can provide only limited insight into the difficulties of synchronizing national with group interests or balancing ideological, political, and economic forces. More empirical evidence is needed before a conceptual reassessment of the development and decline of the regime is undertaken. Focusing on the growing disjuncture amongst the economic, political, and ideological levels in the regime, however, seems to explain why the triad of regime, textile, and colonial interests was not eventually a successful one, and may explain more fully the eventual collapse of the regime. The regime did not completely stagnate during the 1950s and 1960s as some interpretations suggest nor was it simply abandoned by rapidly expanding financial-industrial conglomerates as the integration view maintains; rather, it seemed to have been torn apart by the centrifugal pull of heterogeneous forces. Unable to reconcile dynamic economic change at home, an outdated ideology, and political rigidity, the regime ultimately succumbed to the conflicting tensions in Portuguese society.

The conflicts and contradictions amongst regime, industrial, and colonial interests have left an unfortunate legacy. Democratic Portugal continues to wrestle with the effects of nearly half a century of state favouritism and bureaucratic omnipotence. The majority of textile mills face the daily challenge of either modernizing in order to compete effectively in their own and European markets or collapsing under the burden of debt, high labour turnover, and deteriorating machinery. Finally, thousands of Africans included in the colonial cotton production scheme still cope with the psychological, social, and economic effects of a policy which was brutally enforced and rigorously controlled. For those affected by the system—from textile manufacturers to African producers—the battle with the past continues.

Appendix

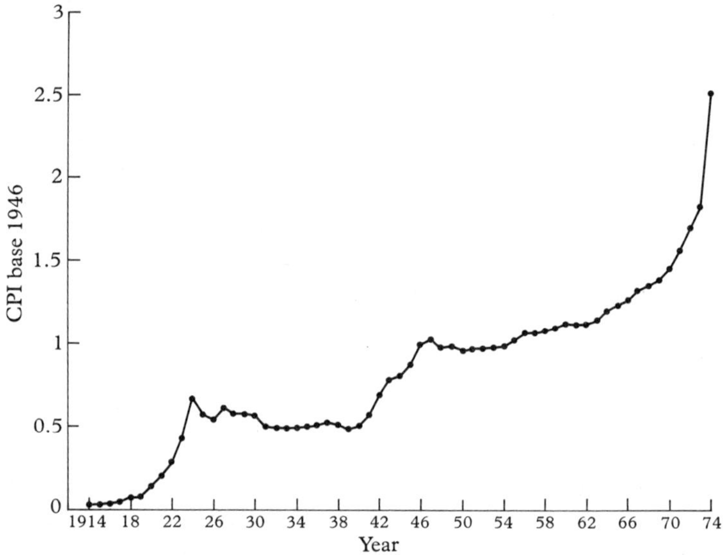

FIG. A1. *Consumer Price Index—Portugal, 1914–1974.*

Note: Retail prices of 25 common consumer goods.
Source: INE, *Anuário Estatístico*, 1914–74.

TABLE A1. *Exchange Rates: Number of Escudos to the Pound Sterling, 1910–1974*

1910	4.90	1927	94.75	1944	100.50	1961	80.83
1911	4.89	1928	98.67	1945	100.50	1962	80.83
1912	4.97	1929	98.75	1946	100.50	1963	80.83
1913	5.25	1930	98.88	1947	100.50	1964	80.83
1914	5.66	1931	106.14	1948	100.50	1965	80.83
1915	6.72	1932	110.03	1949	95.13	1966	80.83
1916	7.01	1933	108.95	1950	80.92	1967	79.53
1917	7.69	1934	110.20	1951	80.92	1968	69.02
1918	7.90	1935	110.20	1952	80.92	1969	68.52
1919	8.21	1936	110.20	1953	80.92	1970	68.85
1920	18.33	1937	110.20	1954	80.92	1971	69.17
1921	39.22	1938	110.19	1955	80.92	1972	67.98
1922	65.39	1939	109.94	1956	80.92	1973	60.26
1923	109.37	1940	102.70	1957	80.92	1974	59.57
1924	133.95	1941	100.50	1958	80.92		
1925	97.38	1942	100.50	1959	80.83		
1926	94.77	1943	100.50	1960	80.83		

Sources: Clarence-Smith, *The third Portuguese empire*, 227–8 (1910–36); Banco de Portugal, *Banco de Dados* (1937–74).

TABLE A2. *Exchange Rates: Number of Escudos to the US Dollar, 1934–1974*

1934	21.87	1945	24.92	1956	28.90	1967	28.96
1935	22.48	1946	24.94	1957	28.90	1968	28.85
1936	22.17	1947	25.02	1958	28.90	1969	28.68
1937	22.29	1948	25.06	1959	28.78	1970	28.74
1938	22.55	1949	26.16	1960	28.79	1971	28.31
1939	24.94	1950	28.95	1961	28.86	1972	27.17
1940	27.56	1951	28.90	1962	28.80	1973	24.63
1941	25.20	1952	28.90	1963	28.87	1974	25.52
1942	25.05	1953	28.90	1964	28.94		
1943	25.05	1954	28.90	1965	28.90		
1944	24.98	1955	28.90	1966	28.94		

Sources: FAO Production Yearbook 1950 (1934–6);
Banco de Portugal, *Banco de Dados* (1937–74).

TABLE A3. *Index of Industrial Plant Growth, 1917–1930 (1917 = 100)*

Industry	1930: Index
Chemicals	895
Food-Processing	529
Beverages	529
Electric	509
Metalworking	388
Ceramics	328
Glassmaking	344
Clothing	286
Wood and Furniture	267
Paper	236
Printing	202
Textiles	189

Sources: Schwartzman, *Social Origins of Democratic Collapse*, 94; and id., 'Lucros, investimentos e coligações políticas na I República', *Análise Social*, 18: 72–4 (1982–3), 741–58.

TABLE A4. *Portugal: Ginned Cotton Imports, 1870–1974* (Unit = Metric tons)

Year	Angola	Moz.	Foreign	Total	Colonies as % of total
1870				1,595	
1880				3,384	
1890				8,232	
1900				15,964	
1910	168*		16,192	16,360	1
1920	404*		11,356	11,760	3
1925	983*		13,481	14,464	7
1926	832	334	15,455	16,621	7
1927	571	292	15,742	16,605	5
1928	520*		16,194	16,714	3
1929				16,448	
1930	641	190	17,137	17,968	5
1931				14,799	
1932	709	1,084	19,589	21,382	8
1933				21,664	
1934	720	1,919	19,009	21,648	12
1935	1,154	1,829	20,458	23,441	13
1936	2,099	3,247	15,175	20,521	26

TABLE A4. *Continued*

Year	Angola	Moz.	Foreign	Total	Colonies as % of total
1937	3,107	8,226	18,860	30,193	38
1938	3,063	7,493	17,789	28,345	37
1939	4,336	6,576	8,253	19,165	57
1940	3,979	4,473	15,530	23,982	35
1941	4,809	5,013	13,280	23,102	43
1942	4,533	14,146	4,361	23,040	81
1943	3,785	13,246	1,014	18,045	94
1944	6,157	22,660	4,594	33,411	86
1945	2,604	17,056	3,888	23,548	93
1946	8,678	29,004	1,803	39,485	95
1947	4,878	17,061	3,924	25,863	85
1948	4,240	24,831	3,752	32,823	89
1949	3,771	25,795	7,917	37,483	79
1950	6,676	25,290	1,471	33,437	96
1951	3,880	23,628	8,670	36,178	76
1952	5,785	28,397	7,929	42,111	81
1953	6,902	31,226	3,850	41,978	91
1954	4,568	38,508	1,227	44,303	97
1955	6,107	31,902	10,607	48,616	78
1956	5,474	21,126	13,208	39,808	67
1957	7,110	22,145	17,939	47,194	62
1958	6,099	35,444	6,457	48,000	87
1959	5,741	29,184	14,169	49,094	71
1960	8,782	39,718	9,036	57,536	84
1961	4,919	42,102	20,286	67,307	70
1962	3,935	29,095	32,222	65,252	51
1963	5,744	36,289	29,986	72,019	42
1964	3,090	28,526	46,601	78,217	40
1965	4,466	36,119	48,864	89,449	45
1966	4,950	27,639	46,881	79,470	41
1967	6,337	37,907	27,603	71,847	62
1968	11,921	41,546	35,780	89,247	60
1969	17,750	45,480	34,392	97,622	65
1970	23,050	44,680	23,986	91,716	74
1971	35,619	34,150	20,662	90,431	77
1972	17,125	26,437	71,264	114,826	38
1973	21,635	33,710	90,651	145,996	38
1974	13,171	26,639	45,571	85,381	47

* Figure includes Angola and Mozambique.

Sources: Bastos, *O algodão no comércio*, 87 (1870–1925); Junta de Exportação de Algodão em Rama (1928); Portugal, INE, *Estatísticas do Comércio Externo* (1926–74).

TABLE A5. *Portugal: Textile Exports to the Colonies and Abroad, 1890–1973 (Unit: Metric Tons)*

Year	Colonies	Abroad	Total	Year	Colonies	Abroad	Total	Year	Colonies	Abroad	Total
1890	130[a]	—	—	1922	1,901	—	2,196	1949	6,300	1,369	7,669
1891	131	—	—	1923	1,334	—	1,360	1950	8,792	5,865	14,657
1892	339	—	—	1924	1,288	—	1,309	1951	10,082	3,722	13,804
1893	693	—	—	1925	1,071	—	1,150	1952	9,605	1,667	11,272
1894	964	—	—	1926	1,046	—	1,046	1953	10,192	2,401	12,593
1895	1,366	—	—	1927	—	—	1,070	1954	10,102	8,759	18,861
1896	1,578	—	—	1928	974	—	1,321	1955	10,622	4,897	15,519
1897	1,834	—	—	1929	1,298	—	1,695	1956	10,966	5,584	16,550
1898	3,893	—	—	1930	—	—	1,029	1957	11,514	7,574	19,088
1899	3,894	—	—	1931	958	—	1,072	1958	11,184	6,987	18,171
1900	3,015	—	3,134	1932	2,071	—	2,378	1959	11,518	6,692	18,210
1901	1,322	—	—	1933	2,609	—	2,947	1960	10,164	19,641	29,805
1902	981	—	—	1934	—	—	2,590	1961	8,907	20,283	29,190
1904	—	—	3,194	1935	2,950	19	2,969	1962	10,450	26,783	37,233
1908	—	—	1,368	1936	2,762	187	2,949	1963	12,882	28,892	41,774
1910	3,864[b]	—	4,281	1937	4,158	2,517	6,675	1964	15,668	29,178	44,846
1911	1,700	—	1,813	1938	2,446	3,114	5,560	1965	15,088	25,756	40,844

1912	1,301	—	1,410	1939	3,251	578	3,829	1966	12,838	36,658	49,496
1913	1,287	—	1,364	1940	2,972	467	3,439	1967	8,292	34,608	42,900
1914	1,115	—	1,193	1941	5,303	4,085	9,388	1968	9,859	30,868	40,727
1915	2,113	—	2,630	1942	4,899	716	5,615	1969	11,061	32,811	43,872
1916	1,937	—	2,078	1943	4,615	440	5,055	1970	7,550	30,668	38,218
1917	1,440	—	1,566	1944	2,926	708	3,634	1971	6,622	30,745	37,367
1918	1,136	—	1,193	1945	5,055	1,662	6,717	1972	3,822	44,013	47,835
1919	828	—	1,079	1946	7,518	428	7,946	1973	4,528	22,861	27,389
1920	2,000	—	2,130	1947	7,863	185	8,048				
1921	1,088[c]	—	1,141	1948	7,747	289	8,036				

[a] Colonial figures from 1890 to 1902 refer to Angola only. Mónica's figure of 339 for 1891 is, I believe, incorrect given her information on customs duties' revisions. I have used Pirio's figure for 1891; otherwise the two writers have the same figures for the period.

[b] *O Trabalho Nacional* gives a figure of 4,281 tons for that year to Angola alone. See *OTN* 7: 73, (Jan. 1926), for 1910.

[c] The figure for 1921 does not agree with Board of Trade figures which give imports of Portuguese textiles of 1,204 tons for Angola alone.

Sources: Mónica, *Artesãos e operários* (1890–1902, except 1891); Pirio, 'Commerce, Industry and Empire'; *BGC* 7: 71 (May 1931), for 1910, 1928, 1929; Clarence-Smith, *The third Portuguese empire* (1911–25); *OTN* 15: 172 (Apr. 1934), for 1931–3; Portugal, INE, *Estatísticas do Comércio Externo* (1900–1934); Rosas, 'Alguns aspectos' (1935–51); Carlos Teixeira Afonso, *Alguns elementos sobre a indústria algodoeira: 1930 a 1965* (1952–65); OECD, *Trade by Commodities*, Series C (1966–73).

TABLE A6. *Textile Imports: Angola and Mozambique, 1890–1937 (Unit: Metric Tons)*

Year	Angola			Mozambique		
	Portugal	Abroad	Total	Portugal	Abroad	Total
1890	15[a]	2,628	2,643	49	1,483	1,532[b]
1900	3,015	—	—	—	—	—
1910	3,851	249	4,100	13	1,288	1,301[b]
1920	1,600	850	2,450	55	1,327	1,382[b]
1921	1,204	60	1,264	—	—	—
1922	1,447	323	1,770	—	—	—
1925	1,059	720	1,779	—	—	1,415[b]
1926	797	800	1,597	—	—	—
1927	832	1,431	2,263	—	—	2,304[b]
1928	874	1,586	2,460	100	2,941	3,041[c]
1929	1,119	1,180	2,299	179	3,849	4,028[c]

1930	592	370	962	—	—	—
1931	528	229	757	—	—	—
1932	1,159	165	1,324	—	—	1,770
1933	1,289	234	1,523	953	1,270	2,223
1935	—	—	—	1,370	1,140	2,510[d]
1936	—	—	—	1,634	3,751	5,385[d]
1937	—	—	—	2,068	3,965	6,033

[a] Statistics on Angolan imports vary widely. I have tried to give the most reliable and/or most plausible available. Mónica gives a figure of 130 tons for 1890.

[b] The Niassa Company is not included in imports up to 1927.

[c] It is not clear which areas are or are not included in the figures for 1928 and 1929, but quite possibly these figures could include the entire colony.

[d] Figures for 1935 and 1936 do not include the Mozambique Company.

Sources: Mónica, *Artesãos e operários*, for Angola, 1890, 1900; Great Britain, Board of Overseas Trade, *Trade Reports*, for Angola, 1921–7, 1930; and Mozambique, 1925, 1927, 1932–3, and 1935; *BGC* 7: 71 (May 1931), for 1890–1920, 1928–9; *OTN* 15: 180 (Dec. 1934), for Angola, 1931–3; and 19: 227–8 (Nov.–Dec. 1938), for Mozambique, 1936–7.

TABLE A7. *Prices of Ginned Cotton, 1914–1974*
(Unit: Escudos per kilogram)

Year	World Price[a] (American middling)	Colonial Price[b] (Portuguese Type 1)	Year	World Price[a] (American middling)	Colonial Price[b] (Portuguese Type 1)
1914	.37	—	1945	14.03	11.00
1915	.30	—	1946	19.20	11.00
1916	.42	—	1947	19.05	12.50
1917	.82	—	1948	17.75	12.50
1918	1.57	—	1949	18.20	13.50
1919	1.52	—	1950	27.23	13.50
1920	3.07	—	1951	27.84	16.30
1921	3.13	—	1952	24.36	16.30
1922	5.48	—	1953	23.78	16.30
1923	11.97	—	1954	24.07	16.30
1924	19.80	—	1955	21.17	18.30
1925	14.73	—	1956	18.27	18.30
1926	9.60	—	1957	18.56	18.30
1927	7.61	—	1958	16.82	18.30
1928	11.12	—	1959	15.95	18.20
1929	10.47	—	1960	17.40	18.30
1930	9.05	—	1961	17.98	18.80
1931	5.68	—	1962	17.40	19.00
1932	4.80	8.00	1963	16.24	22.00
1933	5.59	8.00	1964	16.53	22.00
1934	6.16	7.50	1965	15.95	19.00
1935	5.72	7.50	1966	15.37	19.00
1936	6.38	6.38	1967	16.24	19.00
1937	4.40	4.40	1968	15.66	19.00
1938	4.55	7.50	1969	15.76	19.00
1939	5.50	7.50	1970	18.11	18.05
1940	6.80	7.50	1971	21.22	18.05
1941	10.10	7.50	1972	21.06	—
1942	11.10	10.00	1973	35.92	—
1943	11.38	10.00	1974	25.34	—
1944	12.05	10.00			

[a] For 1934–51 the world price is that for wholesale American middling; after 1951, American middling, c.i.f. Liverpool. Wholesale American prices were used because during and after World War II, individual countries fixed cotton prices through import subsidies and/or customs tariffs. Wholesale prices therefore give a general idea of the price of cotton during these years without taking into consideration specific controls. Prices quoted were US cents/per kg. These were converted to escudos using the exchange rates in Table A2.

[b] Until 1970–1, the colonial cotton price is that for 1st quality cotton, c.i.f. Leixoes. Although not all cotton bought from the colonies was first quality and the classification system was quite rudimentary, it made up the greatest percentage of cotton purchased and has been used as the basis of comparison for that reason. In 1970 Portugal adopted a 7-grade, 10-length classification system. Price quoted is for Grade 1, 15/16″.

Sources: Colónia de Moçambique, Direcção dos Serviços de Agricultura, *Algodão* (1913–33); *BGC* (1932–43); JEAC (1943–50); FAO, *Production Yearbook*, World Prices (1934–74); *Diário do Governo*, Colonial Prices (1951–71).

TABLE A8. *Real Prices of Ginned Cotton, 1914–1974*
(Unit: Escudos per kilogram)

Year	Real World Price	Real Colonial Price	Year	Real World Price	Real Colonial Price
1914	14.73		1944	14.95	12.41
1915	10.71		1945	16.05	12.58
1916	12.19		1946	19.20	11.00
1917	20.11		1947	18.48	12.13
1918	21.35		1948	18.20	12.82
1919	19.10		1949	18.48	13.71
1920	22.15		1950	28.38	14.07
1921	15.25		1951	28.62	16.75
1922	19.34		1952	25.00	16.73
1923	27.71		1953	24.19	16.58
1924	29.71		1954	24.29	16.45
1925	25.64		1955	20.67	17.86
1926	17.80		1956	17.09	17.11
1927	12.46		1957	17.35	17.11
1928	19.29		1958	15.50	16.87
1929	18.34		1959	14.51	16.56
1930	16.06		1960	15.49	16.29
1931	11.36		1961	16.04	16.77
1932	9.80	16.34	1962	15.53	16.96
1933	11.42	16.35	1963	14.13	19.14
1934	12.46	15.17	1964	13.71	18.25
1935	11.49	15.06	1965	12.81	15.26
1936	12.56	12.56	1966	12.04	14.88
1937	8.33	8.33	1967	12.21	14.29
1938	8.89	14.65	1968	11.55	14.01
1939	11.42	15.58	1969	11.32	13.64
1940	13.47	14.86	1970	12.38	12.34
1941	17.72	13.16	1971	13.51	11.49
1942	16.02	14.43	1972	12.37	
1943	14.53	12.76	1973	19.59	
			1974	10.02	

Notes: Real prices calculated using nominal prices in Table A7 and Portugal's Consumer Price Index for the years 1914 to 1974, see Fig. A1.

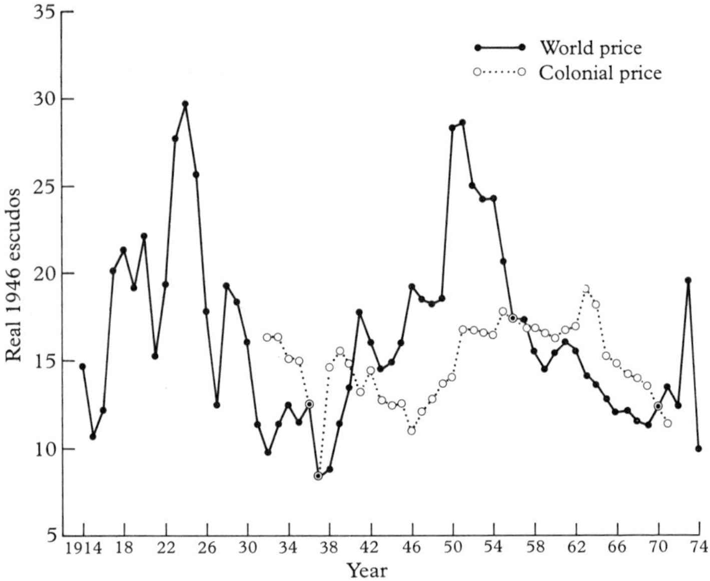

FIG. A2. *Prices of Ginned Cotton, 1914–1974: Real World and Colonial Price.*

TABLE A9. *Average Producer Prices per kilo in Mozambique, 1939–1973* (Unit: Escudos per kilogram)

Year	Quality		Year	Quality	
	1st	2nd		1st	2nd
1939	0.98	0.83	1952	2.70	2.00
1940	0.95	0.85	1953	2.70	2.00
1941	0.95	0.85	1954	2.70	2.00
1942	1.25	1.05	1955	2.70	2.00
1943	1.20	1.00	1956	3.00	2.30
1944	1.20	1.00	1957	3.00	2.30
1945	1.20	1.00	1958	3.00	2.30
1946	1.20	1.00	1959	3.00	2.30
1947	1.50	1.10	1960	3.00	2.30
1948	1.50	1.10	1961	3.60	2.20
1949	1.80	1.40	1963	3.70	2.30
1950	1.80	1.40	1968	3.80	2.20
1951	2.70	2.00	1973	8.00	4.00

Sources: Bravo, *A cultura algodoeira* (1939–61); UEM, CEA, 'Cotton Production in Mozambique' (1963–73).

TABLE A10. *Average Real Producer Prices per kilo in Mozambique, 1939–1973* (Unit: Escudos per kilogram)

Year	Quality		Year	Quality	
	1st	2nd		1st	2nd
1939	2.04	1.72	1952	2.77	2.05
1940	1.88	1.68	1953	2.75	2.03
1941	1.67	1.49	1954	2.72	2.02
1942	1.80	1.52	1955	2.64	1.95
1943	1.53	1.28	1956	2.81	2.15
1944	1.49	1.24	1957	2.80	2.15
1945	1.37	1.14	1958	2.77	2.12
1946	1.20	1.00	1959	2.73	2.09
1947	1.46	1.07	1960	2.67	2.05
1948	1.54	1.13	1961	3.21	1.96
1949	1.83	1.42	1963	3.22	2.00
1950	1.88	1.46	1968	2.80	1.62
1951	2.78	2.06	1973	4.36	2.18

Note: Real producer prices calculated using nominal producer prices in Table A9 and Portugal's CPI for 1939 to 1973, see Fig. A1.

TABLE A11. *Percentages of Unginned Cotton by Quality: Angola and Mozambique, 1939–1946*

Year	Angola			Mozambique		
	1st	2nd	3rd	1st	2nd	3rd
1939	56	35	9	63	22	15
1940	57	30	13	61	17	22
1941	61	37	4	77	11	13
1942	74	21	6	72	10	17
1943	77	17	6	72	10	18
1944	78	14	9	69	7	24
1945	78	13	11	66	9	24
1946	81	9	11	71	6	22

Source: JEAC, *Alguns anos*, 1946.

TABLE A12. *Percentages of Ginned Cotton by Quality: Angola and Mozambique, 1939–1944*

Year	Angola			Mozambique		
	1st	2nd	3rd	1st	2nd	3rd
1939	77	20	3	68	17	15
1940	61	31	7	66	14	19
1941	77	15	8	70	9	21
1942	80	12	8	72	12	16
1943	74	12	14	79	6	15
1944[a]	89	8	4	73	9	10

[a] From 1945, a 6-grade classification was introduced for ginned cotton.

Source: JEAC, *Alguns anos*, 1946.

TABLE A13. *Number of Cotton Textile Plants and Machinery, 1870–1974*

Year	Total			Breakdown by Type of Factory			
	Factories	Spindles Installed	Looms Installed	Spinning only	Spinning and Weaving	Weaving only	Other
1870		50,000					
1881	97[a]	94,167	13,716				
1890		300,000					
1913		480,000					
1928	52	503,000	22,000				
1933	146	550,000	22,000				
1935	232	600,426	22,694				
1940	242	673,873	23,091	22	64	153	3
1941	249	655,822	23,028	22	62	162	3
1942	250	668,135	23,871	22	64	161	3
1943	256	672,029	24,250	22	64	167	3
1944	256	673,949	24,557	22	65	166	3
1945	254	667,309	24,483	22	65	165	2
1946	254	669,147	24,519	22	65	165	2
1947	253	674,545	24,570	22	66	163	2
1948	493[b]	690,927	26,844	21	70	400	2
1949	486	719,884	27,859	22	70	391	3
1950	469	778,896	28,671	22	72	372	3
1951	468	835,940	29,458	21	74	369	4
1955	445[c]	961,242	31,914				
1958	428	1,054,842	32,460	29	76	317	6
1966	436	1,253,357	37,250	33	77	321	5
1970	415[c]	1,357,369	38,740				
1974	396	1,518,415[d]					

[a] Includes cottage mills.
[b] Increase due to Directive 11,879.
[c] A breakdown by factory is unavailable.
[d] Spindle figure is for 1972.

Sources: Mónica, *Artesãos e operários* (1870, 1890); *Inquerito Industrial* (1881); *Indústria Portuguesa* (1913); Taveira, 'Indústria algodoeira' (1928); *OTN* (1933); Rosas, 'Alguns aspectos' (1935–51); II Congresso da Indústria Portuguesa, *Relatório* (1955); Ribeiro, *Reorganização da indústria* (1958); Comissão de Estudo da Indústria Têxtil, *Relatório* (1966); Portugal, INE, *Estatística Industrial* (1970, 1974); Grémio Nacional dos Industrias Têxteis (1970–2), for spindles.

TABLE A14. *Spindles Installed in All Spinning Plants by Category, 1940–1966 (categories of Spindles)*

Year	Total		1–5,000		5,001–10,000		10,001–20,000		20,001–40,000		40,001–60,000	
	Factories	Spindles	Factories	Spindles	Factories	Spindles	Factories	Spindles	Factories	Spindles	Factories	Spindles
1940	89	673,873	47	135,568	18	125,732	18	236,385	6	181,186	—	—
1941	87	655,822	47	136,664	18	128,464	16	212,482	6	181,212	—	—
1942	89	668,135	48	135,672	18	127,921	17	223,230	6	181,312	—	—
1943	89	672,029	48	136,536	18	128,629	17	225,552	6	181,312	—	—
1944	90	673,949	49	139,254	19	138,583	16	214,800	6	181,312	—	—
1945	89	667,309	48	136,132	20	145,173	15	204,692	6	181,312	—	—
1946	89	669,147	48	137,724	20	145,183	15	204,928	6	181,312	—	—
1947	90	674,545	49	142,668	20	145,637	15	204,928	6	181,312	—	—
1948	93	690,927	50	149,178	21	149,855	16	215,792	6	176,552	—	—
1949	95	719,884	50	147,752	20	142,060	18	248,608	6	181,032	—	—
1950	97	778,896	48	140,540	20	138,368	23	314,332	6	185,656	—	—
1951	99	835,940	45	129,162	25	172,714	21	288,516	8	245,548	—	—
1955	104	961,242	43	122,926	27	193,594	24	345,232	9	250,184	1	49,936
1958	111	1,054,842	42	119,260	31	218,954	28	403,428	9	263,264	1	49,936
1966	115	1,253,357	31	85,572	38	264,672	29	407,958	16	436,003	1	59,152

Sources: See Table A13.

TABLE A15. *Looms Installed in All Weaving Plants by Category, 1940–1966 (categories of looms)*

Year	Total		Up to 30		31 to 100		101 to 300		More than 300	
	Factories	Looms	Factories	Looms	Factories	Looms	Factories	Looms	Factories	Looms
1940	217	23,091	91	1,267	73	4,245	31	5,399	22	12,180
1941	224	23,028	96	1,439	77	4,562	31	5,452	20	11,575
1942	225	23,871	95	1,498	77	4,660	32	5,750	21	11,963
1943	231	24,250	94	1,439	84	5,141	32	5,676	21	11,994
1944	231	24,557	92	1,497	84	5,071	34	5,892	21	12,097
1945	230	24,483	90	1,436	86	5,263	33	5,808	21	11,976
1946	230	24,519	91	1,465	84	5,152	34	5,922	21	11,980
1947	229	24,570	90	1,449	82	4,967	36	6,174	21	11,980
1948	470[a]	26,844	324	3,040	88	5,211	36	6,079	22	12,514
1949	461	27,859	313	3,207	91	5,258	34	6,278	23	13,116
1950	444	28,671	290	3,170	93	5,678	37	6,389	24	13,434
1951	443	29,458	287	3,236	95	5,715	37	6,637	24	13,870
1955	414	31,914	244	2,923	100	5,896	45	7,879	25	15,216
1958	396	32,460	225	2,913	100	5,977	45	7,942	26	15,628
1966	401	37,250	196	3,009	108	6,260	64	10,498	33	17,483

[a] Increase due to Directive 11,879.

Sources: See Table A13.

Bibliography

GOVERNMENT AND CORPORATE PUBLICATIONS AND OFFICIAL DOCUMENTS

Afonso, Carlos Teixeira, *Alguns elementos sobre a indústria algodoeira: 1930 a 1965*, Separata dos Bols. Sem. DGSI, 2: 89–101, 1967.

Banco de Portugal, Direcção de Serviços de Estatística e Estudos Económicos, *Banco de Dados*, 21 June 1979.

Comissão de Estudo da Indústria Têxtil, *Relatório e projecto de regulamento de exercício da indústria têxtil*, Oporto, 30 Apr. 1968.

Fonseca, Maria Helena Garcia da, 'Intervencionismo do Estado no sector têxtil algodoeiro', Estudos no. 6, Instituto Nacional de Investigação Industrial, 1962.

Great Britain, Board of Overseas Trade, *Report on the Commercial Situation of Portuguese East Africa*, 1921 (London: HMSO, 1922).

—— —— *Report on Economic Conditions in Portuguese East Africa* (London: HMSO, 1927–56).

—— —— *Report on the Economic Conditions in Angola* (London: HMSO, 1923–53).

Grilo, F. M., 'Sobre a cultura do algodão', Serié: *Divulgação e propaganda* (Bol. no. 1, Direcção dos Serviços de Agricultura; Lourenço Marques: Imprensa Nacional, 1933).

International Labour Organisation, *Portugal in Africa*, Report of ILO Commission on the Observance by Portugal of the Abolition of Forced Labour Convention, 1957 (no. 105), Findings and Recommendations (London: Portuguese Embassy, 1962).

JEAC (Junta de Exportação de Algodão Colonial), *Relatório*, 1942.

—— *Alguns anos de actividade* (Lisbon: Cristovão Augusto Rodrigues, Lda., 1946).

—— *Trabalhos do Centro de Investigação Científica Algodoeira*, i (Lourenço Marques: Minerva Central, 1948).

Moçambique, Direcção dos Serviços de Agricultura, *Algodão*, Colónia de Moçambique, Primeira Exposição Colonial Portuguesa, Oporto (Lourenço Marques: Imprensa Nacional, 1934).

—— Repartição Central de Estatística, *Lista dos exportadores da colónia de Moçambique*, 1937, 1939 (Lourenço Marques: Imprensa Nacional, 1938, 1940).

—— *Relatório da Junta do Credito Agrícola*, 1934–8.

—— *Moçambique Directory*, 51st edn., 1950–1 (Lourenço Marques: A. W. Bayly & Co., Ltd.)

Monteiro, Armindo, *Da governação de Angola* (Lisbon: Agência Geral das Colónias, 1935).

Moreira, Adriano, *A Policy of Integration* (repr. Lisbon: Agência Geral do Ultramar, 1961).

—— *Continuity* (repr. Lisbon: Agência Geral do Ultramar, 1962).

—— *Portugal's Stand in Africa* (repr. New York: University Publishers, 1962).

—— *Revision* (repr. Lisbon: Agência Geral do Ultramar, 1962).

OECD, *Economic Surveys: Portugal* (Paris: OECD, various years).

—— *Trade by Commodities*, Series C, 1967–73.

OEEC, *General Memorandum on the 1950–51 and 1951–52 Programmes: Portugal* (Paris: OEEC, 1950).

—— *Economic Conditions in Member and Associated Countries: Portugal* (Paris: OEEC, 1960).

—— *The Textile Industry in Europe* (Paris: OEEC, various years).

Oliveira, Nunes de, *Moçambique: Alguns aspectos e problemas da sua vida política, económica e social* (Lisbon: Agência Geral das Colónias, 1943).

Pereira, Pedro Teotonio, *Organização corporativa* (Lisbon: Edições SPN, 1935).

Portugal, Agência Geral das Colónias, *Boletim Geral das Colónias*, 1946–51.

—— Agência Geral do Ultramar, *Boletim Geral do Ultramar*, 1952–69.

—— —— *Organic Law of the Portuguese Overseas Provinces* (Lisbon: Agência Geral do Ultramar, 1963).

—— Assembleia Nacional, *II Plano de Fomento (1959–1964)* (3 vols.; Lisbon: Imprensa Nacional, 1959).

—— Câmara Corporativa, Pareceres da Câmara Corporativa, V Legislatura, i (Lisbon: Imprensa Nacional, 1951).

—— *Diário do Governo*, Série I, 1951–74.

—— INE (Instituto Nacional de Estatística), *Estatística Agrícola*, Delegação de Moçambique, Direcção Provincial dos Serviços de Estatística, 1965–7.

—— —— *Anuário Estatístico*, 1925–43.

—— —— *Estatística Industrial*, 1943–72.

—— —— *Estatística da Organização Corporativa*, 1944–70.

—— —— *Estatísticas do Comércio Externo*, 1926–74.

—— —— *I Congresso dos Economistas Portuguesas (problemas das economias ultramarinas)* (Lisbon: Centro de Estudos Económicos, INE, 1955).

—— —— *Inquérito Industrial*, 1957–9, 1964.

—— —— *Recenseamento Industrial*, 1972.

—— Ministério da Economia, *Boletim da Direcção Geral da Indústria*, Publicação Semanal, 1937–49 (formerly Min. do Comércio e Indústria, 1937–9).

—— —— *Boletim da Direcção Geral dos Serviços Industriais*, 1949–65.
—— Ministério das Colónias, *Anuário Colonial*, 1927–9 (Lisbon: Imprensa Nacional, 1930).
—— —— *Conferência dos Governadores Coloniais: Discursos e entrevistas* (Lisbon: Agência Geral das Colónias, 1934).
—— —— *Primeira conferência do Império Colonial Português* (Lisbon: Agência Geral das Colónias, 1934).
—— Ministério das Obras Publicas, Comércio e Indústria, *Inquérito Industrial*, 1881.
—— SPN, *Political Constitution of the Portuguese Republic* (Lisbon: SPN, 1937).
—— Presidência do Conselho, *III Plano de Fomento para 1968–1973*, i (1968).
—— —— *Plano Intercalar de Fomento, 1965–67.*
—— Sub-secretáriado de Estado das Corporações e Previdência Social, *Problemas da organização corporativa*, no. 10, conferência realisada 13 Jan. 1934.
'The Portuguese Constitution', repr. in Amos J. Peaslee, *Constitutions of Nations: Europe*, iii (The Hague: Martinus Nijhoff, 1968).
Ribeiro, José Joaquim Teixeira, *Reorganização da indústria têxtil algodoeira* (Projecto de Relatório da Comissão Reorganizadora, 1966).
Rosas, João Dias, 'Alguns aspectos da situação da indústria algodoeira nacional', Sep. do *Boletim da Direcção Geral dos Serviços Industriais*, 5: 217–23 (8 Apr. 1953).
—— 'Esclarecimento sobre a situação da indústria têxtil algodoeira', Discurso, Vila Nova de Famalicão, 28 Oct. 1957.
Salazar, Antonio de Oliveira, *O momento político* (Lisbon: SPN, 1934).
—— 'Discurso', *I Congresso da União Nacional*, Lisbon, 26–8 May 1934, Vol. I (Lisbon: Edição Nacional, 1935).
—— *Discursos*, 1928–58, 5th edn. (vol. i), 2nd edn. (vols. ii–v) (Coimbra: Coimbra Editora).
—— 'Portugal's Position with Reference to Europe, America, and Africa', Speech before the Chairmen of the District Committee of the National Union, 23 May 1959 (repr. Lisbon: Secretariado Nacional da Informaçao (SNI), 1959).
—— 'Portugal and the Anti-Colonialist Campaign', Speech before the National Assembly, 30 Nov. 1960 (repr. Lisbon: SNI, 1960).
—— 'The Portuguese Overseas Territories and the United Nations Organization', Speech to the Extraordinary Session of the National Assembly held on 30 June 1961 (repr. Lisbon: SNI, 1961).
—— 'Portuguese Problems in Africa', Interview given to *Life* (repr. Lisbon: SNI, 1962).
—— 'Portuguese Overseas Policy', *Africa Institute Bulletin*, 3: 10 (1963), 297–8, 306.

Salazar, Antonio de Oliveira, *Declaration on Overseas Policy*, Broadcast 12 Aug. 1963 (repr. Lisbon: SNI, 1963).
—— 'The Policy in Africa and its Errors', Speech acknowledging the tribute paid to him on 30 Nov. 1967 by the Representatives of the municipalities of Mozambique (repr. Lisbon: SNI, 1967).
Santos, A. C. Valdez dos, *Perspectivas económicas de Angola* (Lisbon: Agência Geral das Colónias, 1949).
UN, FAO, 'World Fibres Review', *Commodity Series*, no. 9, Sept. 1948.
—— —— *Production Yearbook* (Washington, DC: UN FAO, various years).
—— —— '1948–1958: Per Caput Fiber Consumption Levels', *Commodity Bulletin Series*, no. 31, 1960.
—— —— *Commodity Review* (Rome: UN, FAO, various years).
—— Office of Public Information, *A Principle in Torment*, ii. *The United Nations and Portuguese Administered Territories* (New York: UN, 1970).

PRIVATE SECTOR PUBLICATIONS AND DOCUMENTS

Algodoeira Colonial Agrícola, *Relatório*, 1947–71.
Alves, Carlos Garcia, 'Desenvolvimento industrial e problemas de mercados', in *Relatórios de problemas gerais*, II Congresso da Indústria Portuguesa (Lisbon: AIP, 1957).
Associações Industrial Portuguesa e Portuense, 'A fixação de novas indústrias e o Condicionamento Industrial', I Coloquio das Associações Económicas Portuguesas, Luanda, 29 Apr.–2 May (Luanda: Gabinete de Estudos das Associações Económicas, 1963).
Companhia Colonial do Buzi, *Relatório e contas*, 1950–74.
Companhia da Fábrica de Fiação de Tomar, *Balanco, relatório da direcção e parecer do conselho fiscal*, 1943, 1946–74.
Companhia da Zambezia, *Relatório, contas e parecer do conselho fiscal*, 1932–70.
Companhia de Diamantes de Angola, 'Notícia sucinta sobre a sua constituição concessões obtidas e trabalhos realizadas em Angola' (Lisbon: 1929).
Companhia de Fiação e Tecidos de Fafe, *Relatório e contas da direcção e parecer do conselho fiscal*, 1934, 1935, 1937, 1939, 1940, 1945, 1946–7, 1951–74.
—— *Estatuto*, 1946.
Companhia de Fiação e Tecidos de Guimarães, *Relatório da direcção, balanço e parecer do conselho fiscal*, 1932–66.
—— 'Proposta da direcção e conselho fiscal a apresentar em assembleia

geral extraordinária em 3 de Novembro, 1956'.
Companhia de Fomento Colonial, *Relatório, balanço e contas do conselho de administração e parecer do concelho fiscal*, 1939, 1940–1, 1944–7, 1949.
Companhia de Moçambique, *Relatório e contas*, 1929–74.
Companhia dos Algodões de Moçambique, *Relatório*, 1971–3.
Companhia Fabril de Salgueiros SARL, *Relatório de direcção e parecer do conselho fiscal*, 1930–60.
Companhia Fabril do Cávado, *Relatório*, 1915, 1936, 1940–69.
Companhia Fiação de Crestuma, *Relatório*, 1912–68.
—— *Estatutos*, 1946.
Companhia Fiação e Tecidos de Alcobaça, *Relatório da direcção e parecer do conselho fiscal*, 1915, 1916, 1934, 1936–53.
Companhia Fiação e Tecidos do Porto, *Relatório e direcção; parecer do conselho fiscal*, 1912–19, 1921–30, 1932, 1970.
Companhia Fiação Portuense, *Relatório da direccão e parecer do conselho*, 1933–6, 1938.
Companhia Geral dos Algodões de Angola, *Relatório*, 1930–2, 1946–50, 1952, 1954–75.
Companhia Nacional Algodoeira, *Relatório e contas*, 1945–74.
Companhia Nacional de Fiação e Tecidos de Torres Novas, *Relatório*, 1850, 1859, 1906, 1908, 1909, 1936, 1938–60.
Companhia Rio Ave, *Relatório da direcção e parecer do conselho fiscal*, 1911–14, 1918, 1932–71.
—— *Estatutos*, 1946.
Companhia Têxtil do Pungue, *Relatório do conselho de administração*, 1956–60, 1962–6, 1972, 1973.
Corte-Real, C., 'Indústria têxtil algodoeira', in *Discursos, conclusões e estudos sobre a indústria portuguesa*, II Congresso da Indústria Portuguesa (Lisbon: AIP, 1957).
Costa, Ismael Alves da, 'A cultura de algodão em Moçambique', III Congresso Colonial Nacional, 1930.
Cruz, Fernando, 'Política industrial: Algumas reflexões sobre o condicionamento', in *Relatórios de problemas gerais*, II Congresso da Indústria Portuguesa (Lisbon: AIP, 1957).
Empresa Textil Da Cuca, Lda., *Relatório da gerência e parecer do conselho fiscal*, 1958–61.
Fábrica de Fiação e Tecidos de Braga, SARL, *Relatório, balanço e contas da direcção e parecer do conselho fiscal*, 1955–69.
Fábrica de Fiação e Tecidos de Ermezinde, *Relatório e contas da direcção e parecer do conselho fiscal*, 1945, 1953–71.
Fábrica de Fiação e Tecidos de Santo Tirso, Lda., *Relatório e contas da gerência e parecer do conselho fiscal*, 1951, 1954–75.
Fábrica de Fiação e Tecidos do Jacinto, *Relatório*, 1955–73.

Fábrica de Sedas Nogueira, *Relatório*, 1952–69.

FITAVE (Companhia de Fiação e Tecidos de Riba d'Ave, SARL), *Relatório da direcção e parecer do conselho fiscal*, 1959, 1960, 1965.

Graça, Fernando dos Santos, *Companhia Colonial do Buzi* (Lourenço Marques: Imprensa Nacional de Moçambique, 1942).

Grémio Nacional dos Importadores de Algodão em Rama, *Relatório e contas*, 1944, 1947–50.

Indústria Têxtil do Ave, *Relatório*, 1950–69.

Lima, A. A. Lisboa de, 'As colónias portuguesas e a crise económica e financeira de Angola', Desenvolvimento da Conferência realizada a 17 de Julho de 1925 na sede da Sociedade de Geografia de Lisboa (Lisbon: Sociedade de Geografia, 1925).

Moura, Francisco Pereira de, *et al.*, 'Estudos sobre a indústria portuguesa', in *Discursos, conclusões e estudos sobre a indústria portuguesa*, II Congresso da Indústria Portuguesa (Lisbon: AIP, 1957).

Primeiro Congresso da Indústria Portuguesa, *Relatório* (Lisbon: Tip. da Empresa do Anuário Comercial, 1933).

Riopele, *Revista Cultural e de Informação*, nos. 1–5, 1970–1.

Sociedade Agrícola Algodoeira (SAGAL), *Estatutos*, 1945, 1952.

—— *Relatório, balanço e contas do conselho de administração e parecer do conselho fiscal*, 1950–75.

Sociedade Algodoeira de Fomento Colonial, *Relatório, balanço e contas do conselho de administração e parecer do conselho fiscal*, 1950–66.

Sociedade Algodoeira do Zambeze, *Relatório e contas*, 1970–3.

Sociedade de Fiação e Tecelagem Rio Vizela, *Relatório da direcção balanço e parecer do conselho fiscal*, 1953–63.

Sociedade de Fiação e Tecidos Azevedo, Ferreira e Ca, SARL, *Relatório*, 1956–72.

Sociedade de Geografia de Lisboa, *Téses e actas das sessões*, Segundo Congresso Colonial Nacional, 6–10 May 1924 (Lisbon: Tip. e Papel. America, 1924).

Taveira, Henrique Pereira, 'Indústria algodoeira', Conferência, Associação Industrial Portuguesa (Lisbon: Tip. da Ilustração, 1928).

SECONDARY SOURCES

Books

Abshire, David M., 'Minerals, Manufacturing, Power, and Communications', in David M. Abshire and Michael A. Samuels (eds.), *Portuguese Africa: A Handbook* (New York: Praeger, 1969), 294–319.

—— and Samuels, Michael A. (eds.), *Portuguese Africa: A Handbook* (New York: Praeger, 1969).

Almeida, J. A. Simões de, *A indústria algodoeira e a organização corporativa* (Lisbon: Editorial Império, 1940).

—— *A política dos preços do algodão português* (Lisbon: Gráfica Lisbonense, 1943).

Almeida, Pinto de, *A indústria portuguesa e o condicionalismo institucional* (Lisbon: Seara Nova, 1961).

Alvares, Pedro, and Fernandes, Carlos Roma, *Portugal e o Mercado Comun*, i, 2nd edn. (Lisbon: Ed. Portico, 1980).

Amaral, Ferreira do, *Os novos caminhos da indústria portuguesa* (Lisbon: Min. de Economia, 1947).

—— *A industrialização em Portugal* (Lisbon: Tip. da ENP, 1966).

Ameal, João, *Construção do Novo Estado* (Porto: Livraria Tavares Martins, 1938).

Aspin, Chris, *The Cotton Industry* (Aylesbury: Shire Publications, 1981; repr. 1987).

Bailey, Norman, 'The Political Process and Interest Groups', in David M. Abshire and Michael A. Samuels (eds.), *Portuguese Africa: A Handbook* (New York: Praeger, 1969), 146–64.

Baklanoff, Eric N., *The Economic Transformation of Spain and Portugal* (New York: Praeger, 1978).

Baptista, J. do Amparo, *Moçambique: Província portuguesa de ontem e hoje* (Vila Nova de Famalicão: Centro Gráfico, 1962).

Barbosa, Daniel Maria Vieira, *Alguns aspectos da economia portuguesa*, i (Oporto: Livraria Lello e Irmão, 1949).

Bastos, Carlos, *O algodão no comércio e na indústria portuguesa* (Oporto: GNIAR, 1947).

—— *Comércio e indústria têxtil em Portugal* (Oporto: GNIAR, 1950).

—— *Indústria e arte têxtil* (Oporto: Tip. Portugália, 1960).

Bender, Gerald J., *Angola under the Portuguese: The Myth and the Reality* (Berkeley and Los Angeles: Univ. of Calif. Press, 1978).

Bessa, Daniel, and Pimenta, Carlos, 'A economia sob o Estado Novo', in Antonio Costa Pinto *et al.*, *O Estado Novo: Das origens ao fim da autarcia, 1926–1959* (Lisbon: Editorial Fragmentos, 1987), i. 301–9.

Bhagavan, M. R., *Angola's Political Economy 1975–1985* (Research Report no. 75; Uppsala: Scandinavian Institute of African Studies, 1986).

Brandenburg, Frank, 'Development, Finance, and Trade', in David M. Abshire and Michael A. Samuels (eds.), *Portuguese Africa: A Handbook* (New York: Praeger, 1969), 219–52.

Bravo, Nelson Saraiva, *A cultura algodoeira na economia do norte de Moçambique* (Lisbon: Junta de Investigações do Ultramar, 1963).

Brettell, Caroline, *Men who Migrate, Women who Wait* (Princeton, NJ: Princeton Univ. Press, 1986).

Brito, José Maria Brandão de, *A industrialização portuguesa no pós-guerra*

(1948–1965): O Condicionamento Industrial (Lisbon: Publicações Dom Quixote, 1989).

Bruneau, Thomas C., Da Rosa, Victor M. P., and Macleod, Alex (eds.), *Portugal in Development: Emigration, Industrialization, the European Community* (Ottawa: Univ. of Ottawa Press, 1984).

Carlyle-Gall, C. (ed.), *East African Manual* (London: Mining and Industrial Publications of Africa, Ltd., 1929).

Carneiro, Armando, *A grei* (Oporto: GEDES, 1957).

—— *A crise da indústria têxtil algodoeira não e um mito* (Oporto: GEDES, 1958).

—— *Antelóquio de um inquérito económico-social feito a indústria têxtil algodoeira* (Vila Nova de Famalicão: Tipografia Minerva, 1959).

Castro, Armando, *A economia portuguesa do seculo xx (1900–1925)* (Lisbon: Edições 70, 1973).

—— *O sistema colonial português em Africa* (Lisbon: Ed. Caminho, 1980).

Chilcote, Ronald H., *Emerging Nationalism in Portuguese Africa: Documents* (Stanford, Calif.: Hoover Institution Press, 1972).

Clarence-Smith, Gervase, *The third Portuguese empire, 1825–1975: A study in economic imperialism* (Manchester: Manchester Univ. Press, 1985).

Cotta, Freppel, *Economic Planning in Corporative Portugal* (London: P. S. King & Son, Ltd., 1937).

Cruz, Manuel Braga da, *Monárquicos e republicanos no Estado Novo* (Lisbon: Publicações Dom Quixote, 1986).

—— *O Partido e o Estado no Salazarismo* (Lisbon: Ed. Presença, 1988).

Cunha, Silva, *O trabalho indigena* (Lisbon: Agencia Geral das Colónias, 1949).

Ferreira, Eduardo de Sousa, *Portuguese Colonialism from South Africa to Europe* (Freiburg: Aktion Dritte Welt, 1972).

Fields, Rona M., *The Portuguese Revolution and the Armed Forces Movement* (New York: Praeger, 1976).

First, Ruth, *Portugal's Wars in Africa* (Int'l Defense Aid Fund; London: Christian Action Publications Ltd., 1971).

—— *Black Gold: The Mozambican Miner, Proletarian and Peasant* (Sussex: Harvester Press, 1983).

Fonseca, Maria Helena Garcia da, *A unidade económica entre a metropole e o ultramar no direito português* (Lisbon: Gabinete de Investigações Económicas do Instituto Superior de Ciências Económicas e Financeiras, 1961).

Fryer, P., and Pinheiro, Patricia McGowan, *Oldest Ally: A Portrait of Salazar's Portugal* (London: Dobson Books, 1961).

Gaitonde, P. D., *The Liberation of Goa: A Participant's View of History* (London: C. Hurst and Co., 1987).

Gallagher, Tom, *Portugal: A Twentieth-Century Interpretation* (Manchester:

Manchester Univ. Press, 1983).

Galvão, Henrique, and Selvagem, Carlos, *Império ultramarino português*, iii. *Angola* (Lisbon: Empresa Nacional de Publicidade, 1952).

—— *Império ultramarino português*, iv. *Moçambique, India, Macău, Timor* (Lisbon: Empresa Nacional de Publicidade, 1953).

Graham, L. S., and Makler, Harry (eds.), *Contemporary Portugal: The Revolution and its Antecedents* (Austin: Univ. of Texas Press, 1979).

Green, Andrew Wilson, 'Portugal and the African Territories: Economic Implications', in David M. Abshire and Michael A. Samuels (eds.), *Portuguese Africa: A Handbook* (New York: Praeger, 1969), 345–63.

Guerra, Henrique, *Angola, estrutura económica e classes sociais: Os ultimos anos do colonialismo português em Angola* (Lisbon: Edições 70, 1979).

Hammond, R. J., *Portugal's African Problem: Some Economic Facets* (New York: Carnegie Endowment for International Peace, 1962).

—— 'Some Economic Aspects of Portuguese Africa in the Nineteenth and Twentieth Centuries', in Peter Duignan and L. H. Gann (eds.), *Colonialism in Africa*, iv (London: Cambridge Univ. Press, 1975).

Haour, Roger, and Silva, Amado da, *Niveis tecnológicas na indústria têxtil algodoeira portuguesa* (Lisbon: Grupo de Estudos Basicos de Economia Industrial, 1976).

Henderson, Lawrence W., *Angola* (Ithaca, NY: Cornell Univ. Press, 1979).

Henriksen, Thomas H., *Mozambique: A History* (London: Rex Collings, 1978).

Isaacman, Allen F., and Isaacman, Barbara, *Mozambique: From Colonialism to Revolution, 1900–1982* (Boulder, Colo.: Westview Press, 1983).

Kavanagh, Dennis, *British Politics: Continuities and Change* (Oxford: Oxford Univ. Press, 1985).

Keefe, Eugene K., *Area Handbook for Portugal* (Washington, DC: USGPO, 1977).

Lavradio, Marquês do, *Portugal em Africa depois de 1851* (Lisbon: Divisão de Publicações e Biblioteca, Agência Geral das Colónias, 1936).

Leeds, Elizabeth, 'Salazar's "Modelo Economico": The Consequences of Planned Constraint', in Thomas Bruneau *et al.*, *Portugal in Development* (Ottawa: Univ. of Ottawa Press, 1984), 13–51.

Lima, J. D. de Rego, *O Condicionamento Industrial: Instrumento de reorganização* (Lisbon: Tip. da Empresa Nacional de Publicidade, 1966).

Lima, Lino, *A crise da indústria têxtil algodoeira* (Vila Nova de Famalicão: Centro Gráfico, 1967).

Lucena, Manuel de, *A evolução do sistema corporativo português*, i and ii (Lisbon: Perspectivas e Realidades, 1976).

Macedo, Henrique Viega da, *Alguns principios da política social e corporativa portuguesa* (Lisbon: Ministério das Corporações e Previdencia Social, 1958).

Makler, Harry M., 'A Case Study of the Portuguese Business Elite', in Raymond S. Sayers (ed.), *Portugal and Brazil in Transition* (Minneapolis: Univ. of Minn. Press, 1968), 228–41.

—— 'The Portuguese Industrial Elite and its Corporative Relations: A Study of Compartmentalization in an Authoritarian Regime', in L. S. Graham and Harry Makler (eds.), *Contemporary Portugal: The Revolution and its Antecedents* (Austin: Univ. of Texas Press, 1979), 123–65.

Mar, Eli J. E., *Portugal e capital multinacional em Moçambique: 1500–1973*, ii (Kastrup, Denmark: African Studies Editorial, 1975).

Marcum, John, *The Angolan Revolution*, i. *The Anatomy of an Explosion (1950–1962)* and ii. *Exile Politics and Guerrilla Warfare (1962–1976)* (Cambridge, Mass.: MIT Press, 1969 and 1978).

Marques, A. H. de Oliveira, *History of Portugal*, i and ii (New York: Columbia Univ. Press, 1972).

—— *A Primeira República portuguesa* (Lisbon: Livros Horizonte, 1971; 2nd edn. 1975).

Marques, Danilo, *A problemática da actual situação do sector têxtil algodoeiro* (Lisbon: Sociedade Tipográfica, 1967).

Marseille, Jacques, *Empire coloniale et capitalisme française: Histoire d'un divorce* (Paris: Albin Michel, 1984).

Martelli, George, 'The Issues Internationalized', in David M. Abshire and Michael A. Samuels (eds.), *Portuguese Africa: A Handbook* (New York: Praeger, 1969), 367–88.

Martins, Elisio, *Colonialism and Imperialism in Mozambique: The Beginning of the End* (Copenhagen: Akademisk Forlag, 1975).

Martins, Herminio, 'Portugal', in S. J. Woolf (ed.), *European Fascism* (London: Weidenfeld and Nicolson, 1968), 302–36.

—— 'Portugal', in M. S. Archer and Salvador Giner (eds.), *Contemporary Europe: Class, Status and Power* (London: Weidenfeld and Nicolson, 1971), 60–89.

Martins, Maria Belmira, *Sociedades e grupos em Portugal* (Lisbon: Editorial Estampa, 1973).

Mattos, J. Rodrigues de, *Corporativismo em Portugal* (Lisbon: Livraria Moraes, 1937).

—— *Condicionamento Industrial* (Lisbon: Livraria Moraes, 1940).

Mazrui, Ali A., and Tidy, Michael, *Nationalism and New States in Africa* (London: Heinemann, 1984).

Medeiros, Fernando, *A sociedade e a economia portuguesas nas origens do Salazarismo* (Lisbon: A Regra do Jogo, 1978).

Medlicott, W. N., *The Economic Blockade*, i (London: HMSO and Longmans, Green & Co., 1952).

—— *The Economic Blockade*, ii (London: HMSO and Longmans, Green & Co., 1959).

Mondlane, Eduardo, *The Struggle for Mozambique* (Harmondsworth: Penguin, 1969).

Mónica, Maria Filomena, *Artesãos e operários: Indústria, capitalismo, e classe operária em Portugal (1870–1934)* (Lisbon: Instituto de Ciências Sociais, 1986).

Morrison, Rodney J., *Portugal: Revolutionary Change in an Open Economy* (Boston: Auburn House Publishing Co., 1981).

Moura, Francisco Pereira de, *Por onde vai a economia portuguesa?* (Lisbon: Dom Quixote, 1969).

Munro, J. Forbes, *Africa and the International Economy 1800–1960* (London: J. M. Dent & Sons Ltd., 1976).

Munslow, Barry, *Mozambique: The Revolution and its Origins* (London: Longman, 1983).

Newitt, M. D. D., *Portugal in Africa: The Last Hundred Years* (London: C. Hurst & Co., 1981).

Nicholson, Francis, and East, Roger, *From the Six to the Twelve: The Enlargement of the European Communities* (Harlow: Longman, 1987).

Nogueira, Franco, *Salazar*, i (Coimbra: Atlantida Editora, 1977).

Opello, Walter C., *Portugal's Political Development* (Boulder, Colo.: Westview Press, 1985).

Pereira, João Martins, *Indústria, ideologia e quotidiano* (Oporto: Edições Afrontamento, 1974).

—— *Pensar Portugal hoje* (Lisbon: Publicações Dom Quixote, 1971).

Pereira, Miriam Halpern, *A política portuguesa da emigração, 1850–1930* (Lisbon: A Regra do Jogo, 1981).

—— *Assimetrias de crescimento e dependência externa* (Lisbon: Seara Nova, 1974).

Pintado, V. Xavier, *Structure and Growth of the Portuguese Economy* (Geneva: EFTA, 1964).

Pinto, Antonio Costa, *et al.*, *O Estado Novo das origens ao fim da autarcia, 1926–1959*, i (Lisbon: Editorial Fragmentos, 1987).

Pinto, Armando Sevinate, *et al.*, *A agricultura portuguesa no período 1950–1980* (Lisbon: Imprensa Nacional, 1984).

Ponte, Bruno da, *The Last to Leave: Portuguese Colonialism in Africa* (London: International Defence and Aid Fund, 1974).

Raby, D. L., *Fascism and Resistance in Portugal: Communists, Liberals and Military Dissidents in the Opposition to Salazar, 1941–1974* (Manchester: Manchester Univ. Press, 1988).

Rafael, Francisco, *et al.*, *Portugal: Capitalismo e o Estado Novo* (Lisbon: Edições Afrontamento, 1976).

Ribeiro, E. K. de Queiroz, *O algodão em Moçambique* (Beira: Govêrno do

Território da Companhia de Moçambique, 1939).

—— *O algodão: Da colheita à industrialização* (Oporto: Imprensa Portuguesa, 1946).

—— 'A primazia das colónias perante a metropole', Conferência realizada no club dos Ferianos Portuenses em 14 Nov. 1945, a convite da Liga Portuguesa de Profilaxia Social (Oporto: Tip. Mendonça, 1947).

—— *O algodão: Novos processos de produção, comércio e indústria* (Oporto: Oficinas Gráficas de 'O Comércio do Porto', 1965).

Ribeiro, Jeronimo, *Salazar e a economia nacional* (Oporto: Machado e Ribeiro, 1933).

Ribeiro, João Mendes, *A indústria têxtil algodoeira* (Fafe: Tipografia 'A Tradição', 1963).

Robinson, Richard, *Contemporary Portugal: A History* (London: George Allen and Unwin, 1979).

Rosas, Fernando, *O Estado Novo nos anos trinta 1928–1938* (Lisbon: Editorial Estampa, 1986).

—— *Portugal entre a paz e a guerra* (Lisbon: Editorial Estampa, 1990).

Schmitter, Philippe C., *Corporatism and Public Policy in Authoritarian Portugal* (Beverly Hills, Calif.: Sage Publications, 1975).

—— 'The Impact and Meaning of "Non-Competitive, Non-Free and Insignificant" Elections in Authoritarian Portugal', in Guy Hermet, R. Rose, and Alain Rouqué (eds.), *Elections Without Choice* (London: Macmillan, 1978).

—— '"The Regime d'Exception" that became the Rule: Forty-eight Years of Authoritarian Domination in Portugal', L. S. Graham and Harry Makler (eds.), *Contemporary Portugal: The Revolution and its Antecedents* (Austin: Univ. of Texas Press, 1979), 3–46.

Schwartzman, Kathleen, C., *The Social Origins of Democratic Collapse* (Lawrence: Univ. Press of Kansas, 1989).

Serrão, Joel, *A emigração portuguesa* (Lisbon: Livros Horizonte, 1973; 3rd edn., 1977).

Silva, A. E. Duarte, *et al.*, *Salazar e o Salazarismo* (Lisbon: Publicações Dom Quixote, 1989).

Teixeira, Alberto de Almeida, *Angola intangível* (Oporto: Sociedade de Papelaria, 1934).

Telo, Antonio José, *Decadencia e queda da I República portuguesa* (Lisbon: A Regra do Jogo, 1980).

Universidade Eduardo Mondlane, Centro de Estudos Africanos, Instituto de Investigação Cientîfica de Moçambique, *Capital privado, produtividade, e participação—um estudo da fábrica texmoque*, Relatório no. 80/2 (Maputo: UEM, CEA, 1980).

—— —— *Cotton Production in Mozambique: A Survey 1936–1979*, Relatório no. 81/1 (Maputo: UEM, CEA 1981).

Vail, Leroy, and White, Landeg, *Capitalism and Colonialism in Mozambique* (London: Heinemann, 1980).

Wheeler, Douglas L., *Republican Portugal: A Political History 1910–1926* (Madison: Univ. of Wisconsin Press, 1978).

—— and Pelissier, René, *Angola* (London: Pall Mall Press, 1971).

Wiarda, Howard J., *Corporatism and Development* (Amherst: Univ. of Massachusetts Press, 1977).

Wohlgemuth, Patricia, 'The Portuguese Territories and the United Nations', Pamph. *International Conciliation*, no. 545 (Nov. 1963).

Xavier, Alberto Pinheiro, *Portugal e a integração económica Europeia* (Coimbra: Liv. Almedina, 1970).

Periodicals

A Indústria do Norte, 1938–74.

Actividades Económicas, 2: 13 and 16 (1957).

Almeida, Vasco Vieira de, 'Changes in the Structure of the Portuguese Economy', *EFTA Bulletin*, 10: 8 (Nov. 1969), 14.

Amaro, Rogério Roque, 'O Salazarismo na lógica do capitalismo em Portugal', *Análise Social*, 18: 72–4 (1982–3), 995–1011.

Anderson, Perry, 'Portugal and the End of Ultra-Colonialism', *New Left Review*, nos. 15–17 (1962), 83–102.

Bandeira, Antonio Rangel, 'The Portuguese Armed Forces Movement: Historical Antecedents, Professional Demands, and Class Conflict', *Politics and Society*, 6 (1976), 1–56.

Barata, Jose Fernando Nunes, *A devolução de poderes às instituições autónomas não territoriais*, Sep. da *Revista de Direito Administrativo*, 2: 2 e seguinte (Coimbra: Coimbra Editora, Lda., 1958).

—— 'Localização da sede dos organismos de coordenação económica no ultramar', *Estudos de Ciências Políticas e Sociais* (No. 54; Lisbon: Junta de Investigação do Ultramar, 1961).

Blackburn, Robin, 'The Test in Portugal', *New Left Review*, nos. 87–8 (Sept.–Dec. 1974), 5–46.

Brito, José Brandão de, 'Corporativismo e industrialização: Elementos para o estudo do Condicionamento Industrial', *Ler Historia*, no. 6 (1985), 51–60.

—— 'Rumos velhos e novos da industrialização portuguesa', CISEP, 'Evolução recente e perspectivas de transformação de economia portuguesa', Conferência realizado em Lisboa, 27–9 Apr. 1983.

Cabral, Francisco, 'Portugal: Towards an Industrial Society', *EFTA Bulletin*, 13: 4 (May 1972), 7–10.

Cahen, Michel, 'Corporatisme et colonialisme: Approche du cas mozambicain, 1933–1979', Parts I and II. *Cahiers d'etudes africaines*, 23–4: 92–3 (1983), 383–417, and 5–24.

Cain, P. J., and Hopkins, A. G., 'Gentlemanly Capitalism and British Expansion Overseas II: New Imperialism, 1850–1945', *Economic History Review*, 2nd. ser., 40: 1, (Feb. 1987), 1–26.

Cardeira, Armando Gil, 'Alguns aspectos da indústria têxtil algodoeira em Angola', Separata da *Económica Lusitana*, no. 14 (Apr.–June 1959).

Carvalho, Mario de, 'Possibilidade de cultivo de algodões do tipo egípcio em Moçambique', *Boletim da Sociedade de Estudos da Província de Moçambique*, 27: 111 (July–Aug. 1958).

The Economist, 1926.

EFTA Bulletin, 1–15 (1960–74).

Empire Cotton Growing Corporation, *Special Report: Cotton Growing in Southern Africa and the Rhodesias* (London: Billing and Sons, Ltd., 1927).

The Empire Cotton Growing Review, 1– , 1924– , various years.

Folha Têxtil, 1976.

Fortuna, Carlos, 'A mudança de côr do algodão moçambicano: Estado, capital e trabalho no período de entre-guerras', *Revista internacional de estudos africanos*, nos. 10–11 (Jan.–Dec. 1989), 101–25.

Friedland, Elaine A., 'Mozambican Nationalist Resistance: 1920–40', *Transafrican Journal of History*, 8: 1–2 (1979), 117–28.

Grémio Nacional dos Importadores de Algodão em Rama, *Boletim semanal de informações*, nos. 1–50, 1948–9.

—— *Noticiário algodoeiro*, 1961.

Grémio Nacional dos Industriais Têxteis, *Boletim*, 1: 1–2 (1969).

—— *Informação*, Anos I–III, 1972–4.

Head, Judith, 'Underdevelopment and Migrant Labour', *Mozambican Studies*, no. 1 (1980); English edn. no. 1 (1982), 53–71.

Hedges, David, and Rocha, Aurelio, 'A reestruturação do colonialismo em Moçambique, 1938–1944', *Cadernos de História*, 5 (Feb. 1987), 41–73.

—— 'Moçambique durante o apogeu do colonialismo português, 1945–1961', *Cadernos de História*, 6 (Nov. 1987), 29–61.

Indústria Portuguesa, 'Numero especial dedicada aos algodões', 19: 22 (1946).

Indústria Têxtil, 1–3: 1–22 (1936–9).

Isaacman, Allen, 'The Mozambique Cotton Cooperative: The Creation of a Grassroots Alternative to Forced Commodity Production', *African Studies Review*, 25: 2–3 (June–Sept. 1982), 5–25.

—— 'Chiefs, Rural Differentiation and Peasant Protest: The Mozambican Forced Cotton Regime 1938–1961', *African Economic History*, 14 (1985), 15–56.

—— *et al.*, 'Cotton is the Mother of Poverty: Peasant Resistance Against Forced Cotton Cultivation in Mozambique, 1938–1961', *International Journal of African Historical Studies*, 13: 4 (1980), 581–615.

Junta de Investigações do Ultramar, Centro de Estudos Políticos e Sociais,

'Colóquios sobre o II Plano de Fomento (Ultramar)', no. 21 (Lisbon: 1959).

—— *Estudo sobre o absentismo e a instabilidade da mão-de-obra africana*, Estudos de Ciências Políticas e Sociais, 1–3: 20, 35, 44 (Lisbon: Tip. Minerva, 1959–60).

Kirby, A. J., 'The Progress of Cotton Growing in the Tanganyika Territory', *ECGR* 7: 3 (July 1930), 169–80.

Kjolner, Fred, 'Política de algodão', repr. in *Indústria Têxtil*, 3: 22 (Mar.–June 1939), 86.

Leite, Joana Pereira, 'La Reproduction du réseau impérial portugais: Quelques précisions sur la formation du circuit d'or Mozambique/ Portugal, 1959–1973', *Estudos de Economia*, 10: 3 (Apr.–June 1990), 365–401.

Libertas, 'Spotlight on Mozambique', 3: 8 (July 1943).

Lucena, Manuel de, 'Sobre a evolução dos organismos de coordenação económica ligados à lavoura', Part I, *Análise Social*, 14: 56 (1978-4), 817–62.

—— 'Sobre a evolução dos organismos de coordenação económica ligados à lavoura', Part II, *Análise Social*, 15: 57 (1979-1), 117–67.

—— 'Sobre a evolução dos organismos de coordenação económica ligados à lavoura', Part III, *Análise Social*, 15: 58 (1979-2), 287–355.

Macedo, A. C. M., 'Relações entre a indústria nacional e a agricultura colonial', *Boletim da Sociedade de Geografia de Lisboa*, 57a Serié, nos. 5–6 (1939), 267–90.

Martins, Herminio, 'Opposition in Portugal', *Government and Opposition*, 4: 2 (Spring 1969), 250–63.

Martins, Julio, 'Estudo enviou a Comissão de Estudo da Reforma do Condicionamento Industrial', Separata da *Revista Arquitectura Portuguesa e Cerâmica e Edificação* (30 Dec. 1936).

Mónica, Maria Filomena, 'Capitalistas e industriais (1870–1914)', *Análise Social*, 23: 99 (1987-5), 819–63.

Monteiro, Armindo, 'Portugal in Africa', *Journal of the Royal African Society*, 38: 151 (1939), 259–72.

Monteiro, Bartolomeu, 'Alguns problemas da indústria têxtil do nosso pais', *Bol. Ordem Eng.*, Memoria no. 246; 11, 6 (1966), 547–63.

Moura, Francisco Pereira de, 'Estagnação ou crescimento da economia portuguesa?', Sep. da *Revista do Gabinete de Estudos Corporativos*, no. 26 (Apr.–June 1956).

Mozambican Studies, no. 1 (1982).

Neil-Tomlinson, Barry, 'The Nyassa Chartered Company: 1891–1929', *Journal of African History*, 18: 1 (1977), 109–28.

O'Brien, Jay, 'Portugal and Africa: A Dying Imperialism', *Monthly Review*, 26: 1 (May 1974), 19–36.

O'Brien, P. K., 'European Economic Development: The Contribution of the Periphery', *Economic History Review*, 2nd ser., 25: 1 (Feb. 1982), 1–18.

O Seculo, 1926.

O Trabalho Nacional, 1917–18, 1924–38.

Parsons, Revd Clifford J., 'The Torment of Angola', *Africa South in Exile*, 5: 4 (July–Sept. 1961), 72–80.

Patriarca, Fátima, 'O triângulo corporativo: Acta e encenação de um despacho salarial (1946–1947)', *Análise Social*, 23: 99 (1987), 905–44.

Pinto, Antonio Manuel, 'Comunicação', *Indústria do Norte*, 24: 642 (June 1973), 5–12.

Pitcher, M. Anne, 'Sowing the Seeds of Failure: Early Portuguese Cotton Cultivation in Angola and Mozambique, 1820–1926', *Journal of Southern African Studies*, 17: 1 (Mar. 1991), 43–70.

'Portugal in Africa', *Africa Institute Bulletin*, 3: 10 (Nov. 1963), 288–97.

'Portugal in Africa', *British Survey*, no. 158 (May 1962).

Raby, D., 'Portugal, 1942–1947: O problema da estratégia revolucionária da oposição', *Análise Social*, 20: 84 (1984-5), 687–700.

Rebelo, Augusto de Sá Viana, 'Discurso', *A Indústria do Norte*, 123: 635 (Nov. 1972), 16–19.

Reis, Jaime, 'A industrialização num pais de desenvolvimento lento e tardio: Portugal, 1870–1913', *Análise Social*, 23: 96 (1987-2), 207–27.

Ribeiro, J. M. Felix, *et al.*, 'Grande indústria, banca e grupos financeiros (1953–73)', *Análise Social*, 23: 99 (1987), 945–1018.

Rocha, Edgar, 'Colónias e exportação de mão-de-obra como fontes de divisas: Considerações sobre a contribuição dos emigrantes para o subdesenvolvimento económico português', *Análise Social*, 18: 72–4 (1982-3), 1053–75.

Rothkugal, Carrie, 'The Good Neighbour', *Libertas*, 3: 8 (July 1943), 27–9, 37–47.

Schwartzman, Kathleen, 'Lucros, investimentos e coligações políticas na I República', *Análise Social*, 18: 72–4, (1982-3), 741–58.

Smith, Alan K., 'Antonio Salazar and the Reversal of Portuguese Colonial Policy', *Journal of African History*, 15: 4 (1974), 653–67.

Standard Bank Group, *Annual Economic Review: Angola* (1968 and 1972).

Vail, Leroy, and White, Landeg, '"Tawani, Machambero!" Forced Cotton and Rice Growing on the Zambezi', *Journal of African History*, 19: 2 (1978), 239–63.

X, 'Algumas considerações oportunas acerca da fundação do "Grémio dos Importadores de Algodão"', *Indústria Têxtil*, 1: 3 (Jan. 1937), 15.

Unpublished Manuscripts and Dissertations

Fortuna, Carlos, 'Threading Through: Cotton Production, Colonial Mozambique and Semi-Peripheral Portugal in the World Economy', Ph.D. thesis (SUNY-Binghamton, 1989).

Head, Judith, 'Sena Sugar Estates and Migrant Labour', Mozambique, Proceedings of a seminar, Centre of African Studies, Univ. of Edinburgh, 1–2 Dec. 1978.

—— 'State, Capital and Migrant Labour in Zambezia, Mozambique: A Study of the Labour Force of Sena Sugar Estates Limited', Ph.D. thesis (Univ. of Durham, 1980).

Ingerson, Alice, 'Corporatism and Class Consciousness in Northwestern Portugal', Ph.D. thesis (Johns Hopkins, 1984).

Leeds, Elizabeth Rachel, 'Labor Export, Development, and the State: The Political Economy of Portuguese Emigration', Ph.D. thesis (MIT, 1984).

Pacheco, Fernando, 'Agricultura e sociedade rural na Angola dos anos 60: O caso de Malange', II Colóquio Internacional em Ciências Sociais sobre a Africa de Língua Oficial Portuguesa, Guiné-Bissau, 19–26 Nov. 1991.

Pirio, Gregory Roger, 'Commerce, Industry and Empire: The Making of Modern Portuguese Colonialism in Angola and Mozambique, 1890–1914', Ph.D. thesis (UCLA, 1982).

Pitcher, M. Anne, 'The Agrarian Question in Mozambique: Recalling the Past to Give Meaning to the Present', M.Phil. thesis (Oxford, 1985).

Ribeiro, João Mendes, 'Interesses de industrialização no ultramar', Comunicação ao *I Colóquio das Associações Económicas Portuguesas*, 30 Apr. 1963.

Ribeiro, Maria Teresa L., and Girão, José Antonio, 'A indústria têxtil, do vestuário e do sector através de funções de produção', Working Paper no. 29, Universidade Nova de Lisboa, 1985.

Silva, Joaquim Ramos, 'As relações económicas luso-americanas e a adesão de Portugal a CEE', unpub. paper presented at conference of the Instituto de Estudos Estrategicos e Internacionias, Jan. 1989, 28.

Vaz, José Manuel, 'Portugal and Her Overseas Territories: A Study in Economic Association and Co-operation', Ph.D. thesis (Exeter, 1978).

Index